T0233690

CISM COURSES AND LECTURES

The series presents lecture notes, monographs, edited works and proceedings in the field of Mechanics, Engineering, Computer Science and Applied Mathematics.
Purpose of the series is to make known in the international scientific and technical community results obtained in some of the activities organized by CISM, the International Centre for Mechanical Sciences.

INTERNATIONAL CENTRE FOR MECHANICAL SCIENCES

COURSES AND LECTURES - No. 358

STEEL PLATED STRUCTURES

EDITED BY

M. IVÁNYI
TECHNICAL UNIVERSITY OF BUDAPEST
AND
M. ŠKALOUD
CZECH ACADEMY OF SCIENCES

Springer-Verlag Wien GmbH

Le spese di stampa di questo volume sono in parte coperte da
contributi del Consiglio Nazionale delle Ricerche.

This volume contains 238 illustrations

In order to make this volume available as economically and as
rapidly as possible the authors' typescripts have been
reproduced in their original forms. This method unfortunately
has its typographical limitations but it is hoped that they in no
way distract the reader.

ISBN 978-3-211-82742-0   ISBN 978-3-7091-3002-5 (eBook)
DOI 10.1007/978-3-7091-3002-5

PREFACE

New kinds of steel structures have recently been developed with the view to make constructional steelworks more competitive with respect to structures made of other materials. Plated structures (i.e. structures made of more or less thin plate elements) playing a very important role among them. Their successful practical application requires adequate scientific data for reliable design, which has been the goal of numerous research teams worldwide, reporting on a good many aspects in the behaviour of steel plated structures. This new evidence, able to play a very important role in the materialisation of the headway mentioned above, concerns mainly the following areas: (i) methods of theoretical analysis of plated structures, (ii) shear lag and "breathing" in the elements of plated structures, (iii) local effects (connected with the way large forces are introduced into the system concerned), (iv) behaviour of parts (whether these be plate or box beams and columns) of steel frames, (v) fatigue phenomena in plated structures subject to repeated loading, (vi) stability phenomena in plated structures, and (vii) composite (steel-concrete) plated structures.

For these reasons, an International Advanced School on "Steel Plated Structures" was held at the International Centre for Mechanical Sciences in Udine on September 26 to 30, 1994 with the view to give complete information about all aspects of the design of plated structures. Following this aim, the course presented not only complete scientific background, but also established recommendations, procedures and formulae for practical design, profiting also from the fact that over the last years the invited lecturers took part in the preparation of several national and international Design Codes for structural steelwork.

The objective of this monograph is to summarize the most important parts of the lectures presented during the aforementioned International Advanced School and thus substantially to enlarge the very positive results and impact of the School.

The monograph will therefore be of interest for a wide range of readers dealing with research, teaching and design of all kinds of steel structures.

M. Iványi
M. Skaloud

CONTENTS

Page

Preface

Page

Preface

METHODS OF THEORETICAL ANALYSIS
OF PLATED STRUCTURES

V. Krístek

Czech Technical University, Prague, Czech Republic

Abstract

Various methods of analysis of thin-walled structures are presented and mutually compared. Stemming from the classical approaches, which are still valuable because of their clearness, simplicity, analytical formulation, closed form solutions, and possible hand calculations, the more advanced methods, exact or approximate, which take advantages of capabilities of digital computers, are discussed.

Introduction

In building, the use of thin-walled structures in the various types of civil engineering construction has increased with the economic necessity of providing high strength with low weight and cost. Box beams, plated structures and other forms of thin-walled structures have many different structural applications. Possibly the most widely well-known applications of thin-walled construction are found in bridges. The high torsional stiffness of box girders, the favourable pattern of distribution of flexural stresses, and the arising of small shear stresses can be regarded as their greatest advantage. This is why box girders are particularly used for large span bridges, for curved bridges, and also for bridges supported by narrow pillars. The advent of high rise buildings has led to the development of efficient methods of construction which include the use of concrete cores supporting a suspended framework. The core can also be considered as a thin-walled cantilever box beam subjected to axial, bending, shear and torsional loads. There are many other applications of thin-walled structures e.g. crane girders which are only indirectly associated with civil engineering construction, but they also provide an illustration of modern use of thin-walled members and their advantages over traditional structural forms.

A successful practical application of thin-walled structures requires adequate analytical methods for reliable design. Several methods of structural analysis of thin-walled structures are commonly used. The theories presented here provide methods of analysis from simple tools which may be used for analyses of simplified structural arrangements to general methods intended for problems and structures of such complexity which are outside the scope of simple and rapid methods. Clearly it is impracticable to embark upon a series of lengthy, complex and costly analyses to obtain accurate solutions simply for purposes of comparison where a quick analytical tools would suffice. On the other hand, the final design must be supported by the most accurate calculation possible. The following methods of analysis are regarded as suitable analytical tools covering all range of practical requirements.

1. Analysis of Plated Girders as Thin-Walled Beams

In this method, the actual thin-walled space structure is regarded as a single beam. Thin-walled beams are structures that can be treated as one-dimensional, characterized by cross-section deformation variables and cross-section internal forces that depend only on the longitudinal coordinate. The application of the method is advantageous if the girder cross section is not too complicated and if the cross-sectional dimensions are small in relation to the span. The girder may have a variable cross section, the various parts of the structure may be made of materials having different properties, and the structural system may be fairly complex. The analysis is based on a number of simplifying assumptions and belongs therefore into the group of methods based on the ordinary theory of elasticity. The hand calculation suffices in many cases, using a computer for partial routine problems (evaluation of formulae, solution of systems of equations) in combination with current calculations. The theories of this group can be simplified substantially by introducing further complementary assumptions and by assuming a simpler shape or performance of the structure (constant cross section, symmetry, etc.). The simplification can be carried out up to the stage of formulae or calculations utilizing various analogies (like with a beam on elastic foundation).

In this ordinary theory, the stress state of the thin-walled beam can be of axial (tensional or compressional), flexural (accompanied by the effects of shear) or torsional character. The appearance of this or other type of stress state depends on the position of the resultant of the external loading of the beam with regard to the axis of the shear centers, and also on the support conditions. The shear centre is a point in the cross section plane through which the resultant of the shear flows passes when the beam is bent in an arbitrary direction. Evidently, if the cross section has two axes of symmetry, the shear centre is identical with the centre of gravity placed in the intersection of these axes. If the cross section has one axis of symmetry, the shear centre is located on this axis.

The axis of a thin-walled beam is defined as the locus of shear centres of the cross sections. By relating the external loading to this axis, the loading can be resolved into individual components producing the above mentioned fundamental types of stress states. Tension, compression and bending can be solved according to the elementary engineer's theory with reasonable accuracy.

The main problem of the analysis of thin-walled beams is that of torsion. Beams of open and closed cross sections twist and suffer warping, but their structural performance under torsion is quite different.

The theory of thin-walled beams of open cross-sections rests on two basic simplifying assumptions: the cross section is perfectly rigid in its own plane and the shear strains in the middle surface of the wall are neglected.

The beams of closed cross sections (box girders) also exhibit warping of their cross sections. In contrast to thin-walled beams of open cross sections, however, box girders exhibit two further important modes of deformation: strain in the middle surface of the walls and distortion of the cross section shape (Fig. 1.1).

The box girders thus can be divided into two groups according to their behaviour under torsion. The first group includes girders with rigid cross sections in their own plane. The cross section of such a girder does not remain plane when twisted, it is warped, it rotates in its own plane, but the cross section does not exhibit distortion (the projection of the cross section into this plane does not change from its original shape). The second group covers girders which do not satisfy this condition. To include a box girder in the first group the rigidity of the cross section by a sufficient number of stiff diaphragms must be ensured to prevent the distortion of cross sections. The cross sections of girders belonging to the second group are deformed (Fig. 1.1b) since the box girder must transmit through its transverse flexural rigidity the forces, which in the girders of the first group were taken over by the diaphragms. Consequently, the behaviour of the girders of the two groups is entirely different.

Extensive work has been carried out during the last decades to develop the analytical methods belonging to the category of the thin-walled beams analysis. The methods are well-documented in the literature. From the classical solutions of Timoshenko and Gere [1], Vlasov [2], Bleich [3], Umanskij [4], to the more general methods capable to take into account also distortional effects and changes in cross sectional dimensions, Kollbrunner and Hajdin [5], Bažant and El Nimeiri [6], Křístek [7], [8]. Many other references has been surveyed by Maisel [9].

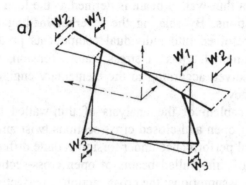

longitudinal warping

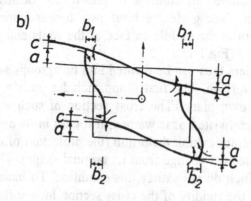

transverse distortion

Fig.1.1. Longitudinal warping and transverse distortion

2. Folded Plate Theory

The folded plate theory is the most exact method of analysis of prismatic thin-walled plated structures. In this method, the thin-walled structure is considered as an assemblage of rectangular plate-shaped parts together forming a real spatial system (Fig.2.1). The bending of each plate element normal to its plane is analysed by plate flexural theory, while in-plane effects are analysed by plane stress theory. Thus, the complexity of the cross section, and the ratio of the cross sectional dimensions to the span, are almost irrelevant. The solution does not make any distinction between open and closed cross sections or multi-celled structures, nor does it differentiate the various kinds of stresses (bending, shear, torsion). The imposed conditions may consist of any combination of loads and displacements. The loading of the structure may be quite general (horizontal, vertical, longitudinal, rotational), arbitrarily distributed on any surface. The solution yields results for displacements and reactions and all of the pertinent internal forces and stresses. Various physical properties of the individual parts are easily introduced into the analysis.

The folded plate theory involves only a relatively few different operations which are repeated many times, it takes full advantage of the capabilities of digital computers and thus it has become a rapid and versatile method for the solution of a very wide range problems of plated structures.

Civil engineering practice makes extensive use of prismatic thin-walled structures of various types. These include bridges, cylindrical shells, roofs and tall buildings, as well as various structural members such as the stiffened compression flanges of steel box girder bridges.

The theory can be used successfully in the analysis of continuous and fixed box girders, frames, girders with interior (right or skew) diaphragms with irregularly situated supports, etc. But even such analysis of more complicated structural systems stems from the solution of the basic problem, viz. a simply supported prismatic folded plate structure without interior diaphragms, framed at both end cross sections into diaphragms infinitely rigid in their own plane and perfectly flexible in the direction perpendicular to this plane, which secure the preservation of the shape of the end cross section, but do not influence longitudinal displacement, that is the warping of these cross sections.

Not only can the overall structural performance be analysed with advantage via the folded plate theory, but this method is also a technique ideally suited to investigations into special problems of considerable impor-tance, such as the problem of shear lag, the problem of stress distribution in structural parts subject to point loads and, particularly, for the stability analysis of plated structures. The folded plate theory offers all these possibilities, thus generally achieving considerable saving of computer costs over other methods.

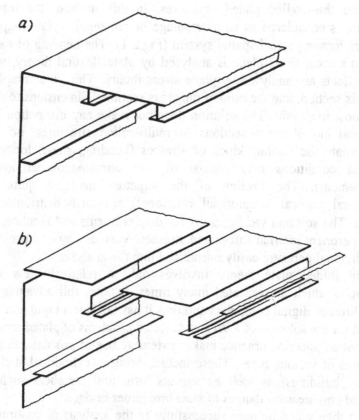

Fig. 2.1. A girder as a system of plate elements.

2.1 Simply Supported Folded Plate Structures with Support Diaphragms

A simply supported prismatic folded plate structure can have the form of, for example, a steel plated girder (Fig. 2.1a), or of a box girder, as shown in Fig. 2.2a. The structure of span L, without interior diaphragms, is framed at both end cross-sections into diaphragms infinitely rigid in their own plane and perfectly flexible in the direction perpendicular to this plane, which thus prevent deformation of the end cross-sections but do not influence longitudinal displacements, that is the warping of these cross-sections. Such a structure may be

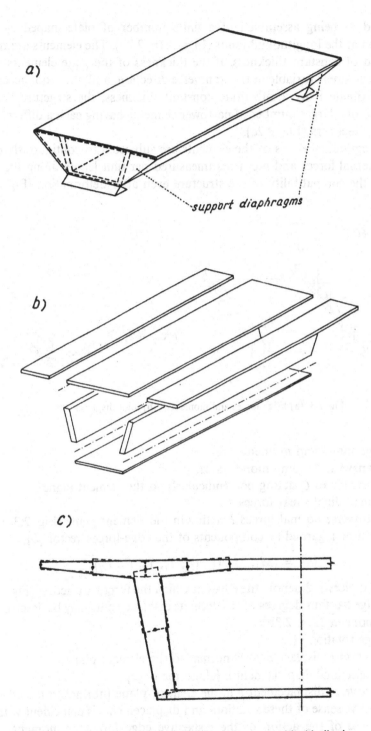

Fig. 2.2. A simply supported folded plate structure and its idealization.

visualized as being assembled of a finite number of plate-shaped elements connected at the longitudinal joints (Figs. 2.1b, 2.2b). The elements are rectangular and of constant thickness. If the thickness of the plate elements of the actual structure is variable in the transverse direction, and if it is not acceptable to approximate it by a substitute constant thickness, the structure is to be composed of a larger number of narrower elements having each a different, but constant, thickness (Fig. 2.2c).

The longitudinal edges of the elements are subjected to continuously distributed internal forces and moments (measured per unit length along the edge) ensuring the compatibility of the structure even after deformation (Fig. 2.3a).

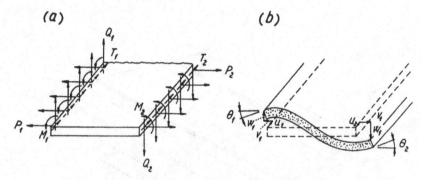

Fig. 2.3. (a) Edge forces and moments, (b) joint displacements.

These edge forces and moments are:
1. transverse bending moments M,
2. shear forces Q acting perpendicularly to the element plane,
3. longitudinal shear forces T,
4. transverse normal forces P acting in the element plane (Fig. 2.3a).

They can be regarded as components of the edge-forces vector $\{S\}$:

$$\{S\} = (M_1, M_2, Q_1, Q_2, T_1, T_2, P_1, P_2)^T,$$

where the indices 1, 2 denote the edges at which the forces are acting (Fig. 2.3a).

Each edge has four degrees of freedom; its displacements may be described by four components (Fig. 2.3b):
1. edge rotation θ,
2. transverse displacement w normal to the element plane,
3. longitudinal displacement u (along the edge),
4. transverse displacement v in the element plane (normal to the edge).

The positive sense of these rotations and displacements is coincident with the positive sense of the action of the respective edge forces or moments. The displacements and rotations of the longitudinal edges are characterized by the edge-displacement vector $\{\delta\}$:

$$\{\delta\} = (\Theta_1, \Theta_2, w_1, w_2, u_1, u_2, v_1, v_2)^{\mathrm{T}}.$$

The relationship between the generalized edge forces $\{S\}$ and the generalized edge displacements $\{\delta\}$ of the respective elements can be written as

$$\{S\} = [k]\{\delta\}, \tag{2.1}$$

where $[k]$ denotes a stiffness matrix of the element, of size 8 by 8. Its form will be derived later.

Thanks to the supports at both ends of the elements placed under the end diaphragms (the diaphragms being rigid in their own plane and perfectly flexible in the direction normal to it), all forces $\{S\}$ and displacements $\{\delta\}$ can be expressed by means of Fourier series (the functions M, Q, P, Θ, w and v in series of the type $\sum_{n=1}^{\infty} \beta_{sn} \sin(n\pi x/L)$, the functions T and u in series of the type $\sum_{n=1}^{\infty} \beta_{cn} \cos(n\pi x/L)$. The n-th term in the series expressing the forces $\{S\}$ produces the displacements $\{\delta\}$, which are again described merely by the n-th term of the corresponding series. This means that Eq. (2.1) can be written independently for each term of the series as follows

$$\{S_n\} = [k_n]\{\delta_n\}. \tag{2.2}$$

Considering that the external loading of the structure can also be expressed by means of Fourier series, the entire analysis can be conducted for each term of the series separately and the final results obtained by summing the partial results. This procedure makes it possible to operate with only the amplitudes of the terms of the series instead of having to deal with functions. The above conclusions also indicate that if the conditions of static equilibrium and geometric compatibility are maintained at a single point of the ridge, they are automatically satisfied along the entire longitudinal joint.

The structural elements being planes, their behaviour can be separated in completely independent slab and membrane actions. The slab action is characterized by the edge moments M and forces Q and by the corresponding joint rotations Θ and transverse displacements w, while the membrane action is given by the edge forces T and P and by the related displacements u and v. This yields a further simplification of the analysis, since the stiffness matrix (of size 8 by 8) of the element can be written in the form

$$[k] = \begin{bmatrix} {}^{d}k & 0 \\ 0 & {}^{s}k \end{bmatrix}, \tag{2.3}$$

where $[{}^{d}k]$ and $[{}^{s}k]$ are the stiffness matrices, of size 4 by 4, which give the slab and membrane actions of the element respectively.

The above relation holds true also for the amplitudes of each harmonic, i.e.,

$$\{S_n\} = \begin{bmatrix} {}^d k_n & 0 \\ 0 & {}^s k_n \end{bmatrix} \{\delta_n\}. \tag{2.4}$$

The stiffness matrices $[{}^d k_n]$ and $[{}^s k_n]$, related to the slab and membrane actions of the element under the loading and deformation given by the n-th harmonic, can be established by means of elasticity theory.

a) Slab action of the element

Let us consider a plate of dimensions L and b. The thickness ${}^d t$, modulus of elasticity ${}^d E$ and Poisson's ratio ${}^d v$ are quantities which can be assumed to differ, in general, from the corresponding constants valid for the membrane action of the element. This approach makes it possible to express, for example, the effect of closely spaced ribs (which increase the flexural stiffness corresponding to the slab action of the elements, whereas the membrane stiffness remains unchanged). or different positions and amounts of reinforcement in the structure, or even various physical properties. The plate is simply supported at two edges (Fig. 2.4a) on the end diaphragms, while the other two edges are free.

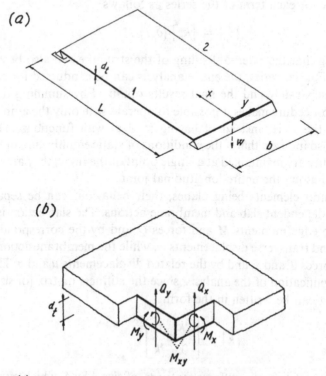

Fig. 2.4. (a) Slab action of an element, (b) slab internal forces and moments.

If the plate is not subjected to any lateral load, the plate deflection described by the function $w(x, y)$ must satisfy the equation

$$\frac{\partial^4 w}{\partial x^4} + 2\frac{\partial^4 w}{\partial x^2 \partial y^2} + \frac{\partial^4 w}{\partial y^4} = 0 . \tag{2.5}$$

For the support of the plate as described, the function w can be written in the following form

$$w(x, y) = \sum_{n=1}^{\infty} A_n(y) \sin \frac{n\pi x}{L} . \tag{2.6}$$

Substituting the n-th term of series (2.6) into Eq. (2.5), one obtains

$$\left(\frac{d^4 A_n}{dy^4} - 2\frac{n^2\pi^2}{L^2}\frac{d^2 A_n}{dy^2} + \frac{n^4\pi^4}{L^4} A_n \right) \sin \frac{n\pi x}{L} = 0 . \tag{2.7}$$

To satisfy relation (2.7), the following differential equation must apply

$$\frac{d^4 A_n}{dy^4} - 2\frac{n^2\pi^2}{L^2}\frac{d^2 A_n}{dy^2} + \frac{n^4\pi^4}{L^4} A_n = 0 . \tag{2.8}$$

Its general solution is

$$A_n(y) = C_{1n} \cosh \frac{n\pi y}{L} + C_{2n}\frac{n\pi y}{L} \sinh \frac{n\pi y}{L} +$$

$$+ C_{3n} \sinh \frac{n\pi y}{L} + C_{4n}\frac{n\pi y}{L} \cosh \frac{n\pi y}{L} , \tag{2.9}$$

$C_{1n}, ..., C_{4n}$ being constants.

In order to derive the stiffness matrix $[{}^d k_n]$ of the element for the plate action, both longitudinal edges of the plate are successively subjected to the deformations Θ and w with unit amplitudes. This means that the plate is to be solved four times, consecutively, with the following boundary conditions:

1.
$$\Theta_{1n}(x) = -\left[\frac{\partial w}{\partial y}\right]_{y=-b/2} = 1. \sin \frac{n\pi x}{L} ,$$

$$\left[\frac{\partial w}{\partial y}\right]_{y=b/2} = w(x, -b/2) = w(x, b/2) = 0 , \tag{2.10}$$

2.
$$\Theta_{2n}(x) = -\left[\frac{\partial w}{\partial y}\right]_{y=b/2} = 1. \sin \frac{n\pi x}{L},$$

$$\left[\frac{\partial w}{\partial y}\right]_{y=-b/2} = w(x, -b/2) = w(x, b/2) = 0, \qquad (2.11)$$

3.
$$w_{1n}(x) = -w(x, -b/2) = 1. \sin \frac{n\pi x}{L},$$

$$\left[\frac{\partial w}{\partial y}\right]_{y=-b/2} = \left[\frac{\partial w}{\partial y}\right]_{y=b/2} = w(x, b/2) = 0, \qquad (2.12)$$

4.
$$w_{2n}(x) = w(x, b/2) = 1. \sin \frac{n\pi x}{L},$$

$$\left[\frac{\partial w}{\partial y}\right]_{y=-b/2} = \left[\frac{\partial w}{\partial y}\right]_{y=b/2} = w(x, -b/2) = 0. \qquad (2.13)$$

Substituting the solution (2.9) into (2.6) and then successively into the boundary conditions (2.10), ..., (2.13), one obtains a system of four linear algebraic equations with four right-hand sides for the constants $C_{1n}, ..., C_{4n}$ in relation (2.9)

$$[F_n]\{C_{in}\} + \{P_{jn}\} = 0, \qquad i, j = 1, ..., 4 \qquad (2.14)$$

the matrix $[F_n]$ having the following form

$$[F_n] = \begin{bmatrix} \sinh \alpha, & \sinh \alpha + \alpha \cosh \alpha, & -\cosh \alpha, & -\cosh \alpha - \alpha \sinh \alpha \\ -\sinh \alpha, & -\sinh \alpha - \alpha \cosh \alpha, & -\cosh \alpha, & -\cosh \alpha - \alpha \sinh \alpha \\ -\cosh \alpha, & -\alpha \sinh \alpha, & \sinh \alpha, & \alpha \cosh \alpha \\ \cosh \alpha, & \alpha \sinh \alpha, & \sinh \alpha, & \alpha \cosh \alpha \end{bmatrix},$$

$$(2.15)$$

where

$$\alpha = \frac{n\pi b}{2L}.$$

The vectors $\{P_{jn}\}$ for the individual cases of the boundary conditions are

$$\left\{\begin{matrix} -1 \\ 0 \\ 0 \\ 0 \end{matrix}\right\}, \left\{\begin{matrix} 0 \\ -1 \\ 0 \\ 0 \end{matrix}\right\}, \left\{\begin{matrix} 0 \\ 0 \\ -1 \\ 0 \end{matrix}\right\}, \left\{\begin{matrix} 0 \\ 0 \\ 0 \\ -1 \end{matrix}\right\}. \tag{2.16}$$

The constants $C_{1n}, ..., C_{4n}$, obtained by solving systems (2.14), substituted into (2.9) and further into (2.6), make it possible to determine the state of stress of the plate under the above mentioned unit deformations according to the relations

$$M_{xn} = -D\left(\frac{\partial^2 w}{\partial x^2} + {}^d v \frac{\partial^2 w}{\partial y^2}\right),$$

$$M_{yn} = -D\left(\frac{\partial^2 w}{\partial y^2} + {}^d v \frac{\partial^2 w}{\partial x^2}\right),$$

$$M_{xyn} = -D(1 - {}^d v)\frac{\partial^2 w}{\partial x\, \partial y},$$

$$Q_{xn} = -D\left(\frac{\partial^3 w}{\partial x^3} + \frac{\partial^3 w}{\partial x\, \partial y^2}\right),$$

$$Q_{yn} = -D\left(\frac{\partial^3 w}{\partial y^3} + \frac{\partial^3 w}{\partial x^2\, \partial y}\right), \tag{2.17}$$

where M_{xn}, M_{yn} are the bending moments (per unit length) in the plate (Fig. 2.4b),

M_{xyn} is the torsional moment,

Q_{xn}, Q_{yn} are the shear forces, and where

$$D = \frac{{}^d E\, {}^d t^3}{12(1 - {}^d v^2)}. \tag{2.18}$$

Applying these constants, it is also possible to determine the reactive moments and the forces acting at the longitudinal plate edges, which are subjected to the deformations according to boundary conditions (2.10), ..., (2.13), whose amplitudes are already elements of the stiffness matrix $\left[{}^dk_n\right]$ for slab action related to the n-th harmonic. While denoting

$$
\left[{}^dk_n\right] = \begin{bmatrix}
{}^dk_{11n}, & {}^dk_{12n}, & {}^dk_{13n}, & {}^dk_{14n} \\
{}^dk_{21n}, & {}^dk_{22n}, & {}^dk_{23n}, & {}^dk_{24n} \\
{}^dk_{31n}, & {}^dk_{32n}, & {}^dk_{33n}, & {}^dk_{34n} \\
{}^dk_{41n}, & {}^dk_{42n}, & {}^dk_{43n}, & {}^dk_{44n}
\end{bmatrix},
\tag{2.19}
$$

the elements of this stiffness matrix are determined by the following relations

$$
{}^dk_{11n} = {}^dk_{22n} = D\,\frac{n\pi}{L}\left(\frac{2\cosh^2\alpha}{2\alpha + \sinh 2\alpha} - \frac{2\sinh^2\alpha}{2\alpha - \sinh 2\alpha}\right),
$$

$$
{}^dk_{12n} = {}^dk_{21n} = -D\,\frac{n\pi}{L}\left(\frac{2\cosh^2\alpha}{2\alpha + \sinh 2\alpha} + \frac{2\sinh^2\alpha}{2\alpha - \sinh 2\alpha}\right),
$$

$$
{}^dk_{13n} = {}^dk_{31n} = D\,\frac{n^2\pi^2}{L^2}\left(\frac{\sinh 2\alpha}{2\alpha + \sinh 2\alpha} - \frac{\sinh 2\alpha}{2\alpha - \sinh 2\alpha} - (1 - {}^d\nu)\right),
$$

$$
{}^dk_{14n} = {}^dk_{41n} = -D\,\frac{n^2\pi^2}{L^2}\left(\frac{\sinh 2\alpha}{2\alpha + \sinh 2\alpha} + \frac{\sinh 2\alpha}{2\alpha - \sinh 2\alpha}\right),
$$

$$
\tag{2.20}
$$

$$
{}^dk_{23n} = {}^dk_{32n} = {}^dk_{14n},
$$

$$
{}^dk_{24n} = {}^dk_{42n} = {}^dk_{13n},
$$

$$
{}^dk_{33n} = {}^dk_{44n} = D\,\frac{n^3\pi^3}{L^3}\left(\frac{2\sinh^2\alpha}{2\alpha + \sinh 2\alpha} - \frac{2\cosh^2\alpha}{2\alpha - \sinh 2\alpha}\right),
$$

$$
{}^dk_{34n} = {}^dk_{43n} = -D\,\frac{n^3\pi^3}{L^3}\left(\frac{2\sinh^2\alpha}{2\alpha + \sinh 2\alpha} + \frac{2\cosh^2\alpha}{2\alpha - \sinh 2\alpha}\right).
$$

b) Membrane action of the element

Let us study a membrane of thickness st, modulus of elasticity ${}^sE$, Poisson's ratio ${}^s\nu$ and dimensions L, b. The supports on the transverse edges are such that only the shear stresses τ_{xy} can develop, whereas the longitudinal normal stresses

σ_x in the support line vanish (Fig. 2.5). This kind of support of the membrane corresponds to support of the element on the end diaphragm of the properties described.

The state of stress of the membrane is characterized by Airy's stress function $\varphi(x, y)$ which must satisfy the equation

$$\frac{\partial^4 \varphi}{\partial x^4} + 2 \frac{\partial^4 \varphi}{\partial x^2 \, \partial y^2} + \frac{\partial^4 \varphi}{\partial y^4} = 0. \tag{2.21}$$

The function φ can be assumed in the following form

$$\varphi(x, y) = \sum_{n=1}^{\infty} B_n(y) \sin \frac{n\pi x}{L}. \tag{2.22}$$

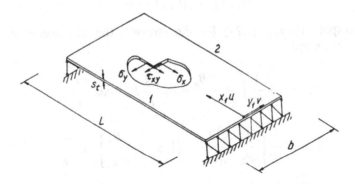

Fig. 2.5. Membrane action of an element.

With due regard to the analogy with relations (2.5) and (2.6) given above for the plate action, the following relation can be written for the coefficients B_n of series (2.22)

$$B_n = K_{1n} \cosh \frac{n\pi y}{L} + K_{2n} \frac{n\pi y}{L} \sinh \frac{n\pi y}{L} + K_{3n} \sinh \frac{n\pi y}{L} + K_{4n} \frac{n\pi y}{L} \cosh \frac{n\pi y}{L},$$

$$\tag{2.23}$$

which corresponds with relation (2.9).

The membrane stresses then read

$$\sigma_{xn} = \frac{\partial^2 \varphi}{\partial y^2} = \frac{d^2 B_n}{dy^2} \sin \frac{n\pi x}{L},$$

$$\sigma_{yn} = \frac{\partial^2 \varphi}{\partial x^2} = -B_n \frac{n^2 \pi^2}{L^2} \sin \frac{n\pi x}{L}, \qquad (2.24)$$

$$\tau_{xyn} = -\frac{\partial^2 \varphi}{\partial x \, \partial y} = -\frac{dB_n}{dy} \frac{n\pi}{L} \cos \frac{n\pi x}{L}.$$

Considering the boundary conditions on the transverse edges of the membrane

$$\sigma_x(0, y) = \sigma_x(L, y) = {}^s\!E \left[\frac{\partial u}{\partial x}\right]_{x=0} = {}^s\!E \left[\frac{\partial u}{\partial x}\right]_{x=L} = 0, \qquad (2.25)$$

$$v(0, y) = v(L, y) = 0$$

and with respect to Eq. (2.24) for the stresses, the displacements of the membrane points are

$$u_n = -\frac{1}{{}^s\!E} \left(\frac{L}{n\pi} \frac{d^2 B_n}{dy^2} + {}^s\!v \frac{n\pi}{L} B_n\right) \cos \frac{n\pi x}{L},$$

$$v_n = \frac{1}{{}^s\!E} \left[\frac{L^2}{n^2 \pi^2} \frac{d^3 B_n}{dy^3} - (2 + {}^s\!v) \frac{dB_n}{dy}\right] \sin \frac{n\pi x}{L}. \qquad (2.26)$$

In order to establish the stiffness matrix of the element for membrane action, both longitudinal edges of the membrane are successively subjected to sinusoidal or cosinusoidal displacement u and v, of unit amplitude. This means that values of constants $K_{1n}, \ldots, K_{4n}$ in relation (2.23) must be found, such that displacements (2.26) satisfy successively four combinations of the boundary conditions

1.
$$u_{1n}(x) = u\left(x, -\frac{b}{2}\right) = 1 \cdot \cos \frac{n\pi x}{L},$$

$$u\left(x, \frac{b}{2}\right) = v\left(x, -\frac{b}{2}\right) = v\left(x, \frac{b}{2}\right) = 0, \qquad (2.27)$$

2.
$$u_{2n}(x) = u\left(x, \frac{b}{2}\right) = 1 \cdot \cos\frac{n\pi x}{L},$$

$$u\left(x, -\frac{b}{2}\right) = v\left(x, -\frac{b}{2}\right) = v\left(x, \frac{b}{2}\right) = 0, \qquad (2.28)$$

3.
$$v_{1n}(x) = -v\left(x, -\frac{b}{2}\right) = 1 \cdot \sin\frac{n\pi x}{L},$$

$$u\left(x, -\frac{b}{2}\right) = u\left(x, \frac{b}{2}\right) = v\left(x, \frac{b}{2}\right) = 0, \qquad (2.29)$$

4.
$$v_{2n}(x) = v\left(x, \frac{b}{2}\right) = 1 \cdot \sin\frac{n\pi x}{L},$$

$$u\left(x, -\frac{b}{2}\right) = u\left(x, \frac{b}{2}\right) = v\left(x, -\frac{b}{2}\right) = 0. \qquad (2.30)$$

Substituting solution (2.23) into relation (2.22) and then successively into boundary conditions (2.27), ..., (2.30), one obtains four linear algebraic equations for the constants K_{1n}, ..., K_{4nn} for each case of boundary deformations. These systems of equations are fully analogous to systems (2.14).

The constants K_{1n}, ..., K_{4n} obtained in the solution of these systems and substituted into (2.23) and further into (2.24), make it possible to derive an expression of the stress state of the membrane under the unit deformations introduced earlier, and also the stresses on the longitudinal edges of the element. These stresses, multiplied by the thickness of the element, determine the reactive forces. The amplitudes of these forces for the n-th harmonic are given by the stiffness matrix

$$[{}^s k_n] = \begin{bmatrix} {}^s k_{11n}, & {}^s k_{12n}, & {}^s k_{13n}, & {}^s k_{14n} \\ {}^s k_{21n}, & {}^s k_{22n}, & {}^s k_{23n}, & {}^s k_{24n} \\ {}^s k_{31n}, & {}^s k_{32n}, & {}^s k_{33n} & {}^s k_{34n} \\ {}^s k_{41n}, & {}^s k_{42}, & {}^s k_{43n}, & {}^s k_{44n} \end{bmatrix}. \qquad (2.31)$$

The elements of the stiffness matrix (2.31) are

$${}^{s}k_{11n} = {}^{s}k_{22n} =$$

$$= 2\frac{{}^{s}E^{s}t}{(1 + {}^{s}v)^{2}} \frac{n\pi}{L} \left(-\frac{\sinh^{2}\alpha}{2\alpha - \dfrac{3 - {}^{s}v}{1 + {}^{s}v}\sinh 2\alpha} + \frac{\cosh^{2}\alpha}{2\alpha + \dfrac{3 - {}^{s}v}{1 + {}^{s}v}\sinh 2\alpha} \right),$$

$${}^{s}k_{12n} = {}^{s}k_{21n} =$$

$$= -2\frac{{}^{s}E^{s}t}{(1 + {}^{s}v)^{2}} \frac{n\pi}{L} \left(\frac{\sinh^{2}\alpha}{2\alpha - \dfrac{3 - {}^{s}v}{1 + {}^{s}v}\sinh 2\alpha} + \frac{\cosh^{2}\alpha}{2\alpha + \dfrac{3 - {}^{s}v}{1 + {}^{s}v}\sinh 2\alpha} \right),$$

$${}^{s}k_{13n} = {}^{s}k_{31n} =$$

$$= -\frac{{}^{s}E^{s}t}{(1 + {}^{s}v)^{2}} \frac{n\pi}{L} \left(\frac{\sin 2\alpha}{2\alpha - \dfrac{3 - {}^{s}v}{1 + {}^{s}v}\sinh 2\alpha} - \frac{\sinh 2\alpha}{2\alpha + \dfrac{3 - {}^{s}v}{1 + {}^{s}v}\sinh 2\alpha} + 1 + {}^{s}v \right),$$

$${}^{s}k_{14n} = {}^{s}k_{41n} =$$

$$= -\frac{{}^{s}E^{s}t}{(1 + {}^{s}v)^{2}} \frac{n\pi}{L} \left(\frac{\sinh 2\alpha}{2\alpha - \dfrac{3 - {}^{s}v}{1 + {}^{s}v}\sinh 2\alpha} + \frac{\sinh 2\alpha}{2\alpha + \dfrac{3 - {}^{s}v}{1 + {}^{s}v}\sinh 2\alpha} \right),$$

$${}^{s}k_{23n} = {}^{s}k_{32n} = {}^{s}k_{14n}, \qquad {}^{s}k_{24n} = {}^{s}k_{42n} = {}^{s}k_{13n},$$

$${}^{s}k_{33n} = {}^{s}k_{44n} =$$

$$= 2\frac{{}^{s}E^{s}t}{(1 + {}^{s}v)^{2}} \frac{n\pi}{L} \left(-\frac{\cosh^{2}\alpha}{2\alpha - \dfrac{3 - {}^{s}v}{1 + {}^{s}v}\sinh 2\alpha} + \frac{\sinh^{2}\alpha}{2\alpha + \dfrac{3 - {}^{s}v}{1 + {}^{s}v}\sinh 2\alpha} \right).$$

$${}^{s}k_{34n} = {}^{s}k_{43n} =$$

$$= -2\frac{{}^{s}E^{s}t}{(1 + {}^{s}v)^2}\frac{n\pi}{L}\left(\frac{\cosh^2\alpha}{2\alpha - \dfrac{3 - {}^{s}v}{1 + {}^{s}v}\sinh 2\alpha} + \frac{\sinh^2\alpha}{2\alpha + \dfrac{3 - {}^{s}v}{1 + {}^{s}v}\sinh 2\alpha}\right).$$

$$(2.32)$$

The element stiffness, characterized by the stiffness matrices corresponding to the slab and membrane actions of the element, is represented by the coefficients given by Eqs. (2.20) and (2.32). These formulae make it possible to take into account the plate thickness and the material characteristics E and v, which may be different for the slab and membrane actions.

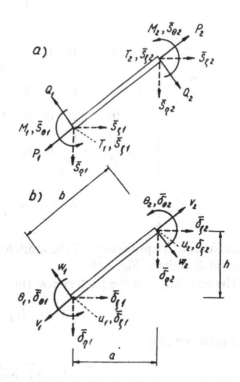

Fig. 2.6. Edge forces and displacements in the local and fixed coordinate systems.

In a folded plate structure, the element is placed in a random position (Fig. 2.6a). For this reason, it is necessary to transform the edge forces $\{S\}$ and displacements $\{\delta\}$ into a fixed coordinate system ξ, η, ζ, valid for the entire folded plate structure.

For the deformation $\{\bar\delta\}$ in the fixed system (Fig. 2.6b)

$$\{\bar\delta\} = (\bar\delta_{\zeta 1}, \bar\delta_{\eta 1}, \bar\delta_{\xi 1}, \bar\delta_{\Theta 1}, \bar\delta_{\zeta 2}, \bar\delta_{\eta 2}, \bar\delta_{\xi 2}, \bar\delta_{\Theta 2})^T, \tag{2.33}$$

it holds true that $[\delta] = [H]\{\bar\delta\}$ and, consequently, also that

$$\{\delta_n\} = [H]\{\bar\delta_n\}, \tag{2.34}$$

where the transformation matrix $[H]$ is

$$[H] = \begin{bmatrix} 0, & 0, & 0, & 1, & 0, & 0, & 0, & 0, \\ 0, & 0, & 0, & 0, & 0, & 0, & 0, & 1 \\ -\dfrac{h}{b}, & -\dfrac{a}{b}, & 0, & 0, & 0, & 0, & 0, & 0, \\ 0, & 0, & 0, & 0, & \dfrac{h}{b}, & \dfrac{a}{b}, & 0, & 0, \\ 0, & 0, & 1, & 0, & 0, & 0, & 0, & 0, \\ 0, & 0, & 0, & 0, & 0, & 0, & 1, & 0 \\ -\dfrac{a}{b}, & \dfrac{h}{b}, & 0, & 0, & 0, & 0, & 0, & 0 \\ 0, & 0, & 0, & 0, & \dfrac{a}{b}, & -\dfrac{h}{b}, & 0, & 0, \end{bmatrix}, \tag{2.35}$$

a and h being the projections of the width b of the element in the horizontal and vertical directions (Fig. 2.6b).

The edge forces in the fixed system (Fig. 2.6b) are

$$\{\bar S\} = [H]^T\{S\}$$

and therefore also

$$\{\bar S_n\} = [H]^T\{S_n\}. \tag{2.36}$$

Combining relations (2.2), (2.34) and (2.35),

$$\{\bar S_n\} = [H]^T[k_n][H]\{\bar\delta_n\} \tag{2.37}$$

or in an aggregate form

$$\{\bar{S}_n\} = [\bar{k}_n]\{\bar{\delta}_n\},\tag{2.38}$$

$[\bar{k}_n]$ being the stiffness matrix of the element related to the n-th harmonic in the fixed coordinate system.

It is very convenient to base the analysis of a folded plate structure on the stiffness method, which is analogous to the procedure frequently used in the analysis of frames. To start with, it is assumed that the structure is fixed along all longitudinal edges of the individual elements, this preventing any displacement or rotation of the joints. The elements are, at this stage, simply supported on the two transverse edges by the end diaphragms and fixed at the longitudinal edges. The stress state produced by the load located between the edges is calculated, and the reactions $\{S_0\}$, acting on the longitudinal fixed edges, determined. The reaction vector

$$\{S_0\} = (\bar{M}_{1,0}, \bar{M}_{2,0}, \bar{Q}_{1,0}, \bar{Q}_{2,0}, \bar{T}_{1,0}, \bar{T}_{2,0}, \bar{P}_{1,0}, \bar{P}_{2,0})^T\tag{2.39}$$

can be established in either of two ways: (a) by directly analysing the element as clamped along the longitudinal edges, for example for the plate action of an element loaded perpendicularly to the plate plane by a load $^d p(x, y)$, Eq. (2.5) is solved with the right-hand side $^d p/D$ and with the boundary conditions $\theta_1 = = \theta_2 = w_1 = w_2 = 0$, or (b) from the influence functions, which are proportional to the displacements $\bar{A}(x, y)$ of the element points, if the longitudinal edges of the elements are subjected to deformations with unit amplitudes, i.e.,

$$1 \sin\frac{n\pi x}{L} \text{ or } 1 \cos\frac{n\pi x}{L} \qquad \text{(see boundary conditions (2.10). ..., (2.13) and}$$

(2.27), ..., (2.30)).

Let us consider a virtual deformation of the element edge $\bar{1}. \sin m\pi x/L$; then only one reaction $S_0(x)$(Fig. 2.3), corresponding to this deformation, performs work at this edge. It follows that

$$\int_0^L S_0(x) \sin\frac{m\pi x}{L} \, dx + \iint_f p(x, y) \bar{A}(x, y) \, dx \, dy + \sum_{j=1}^N P_j \bar{A}(x_j, y_j) = 0,\tag{2.40}$$

where f is the loaded surface of the element,

P_j is the force loading the element,

$\bar{A}$ is the displacement of the element points produced by the deformation of the edge.

Assuming the reaction in the form of a series, we can write

$$S_0(x) = \sum_{n=1}^{\infty} s_n \sin \frac{n\pi x}{L}.$$ (2.41)

Relation (2.40) then holds for the n-th term of the series:

$$s_n \int_0^L \sin \frac{n\pi x}{L} \sin \frac{m\pi x}{L} \, dx + \iint_f p(x, y) \, \bar{A}(x, y) \, dx \, dy + \sum_{j=1}^{N} P_j \bar{A}(x_j, y_j) = 0.$$

(2.42)

The value of the first integral in relation (2.42) differs from zero only for $m = n$, when it equals $L/2$. Thus

$$s_n = -\frac{2}{L}\left[\iint_f p(x, y) \, \bar{A}(x, y) \, dx \, dy + \sum_{j=1}^{N} P_j \bar{A}(x_j, y_j) \right].$$ (2.43)

A similar procedure could be used for the virtual deformation of the longitudinal edge of the element in the form of $\bar{1}. \cos(m\pi x/L)$ and for the reaction

$$S_0(x) = \sum_{n=1}^{\infty} s_n \cos(n\pi x/L).$$

It follows from relation (2.43) that the displacements $\bar{A}$ are equal to the $L/2$-fold influence functions of the amplitude of the n-th term of the series (2.41), these functions expressing the reaction which acts in the direction of the forced deformation.

The reactions $\{S_0\}$ should be transformed into the global coordinate system.

To carry out an analysis of the folded plate structure by the stifffness method, the fixation in the joints (which was assumed for the determination of the reactions of the load acting between the ridges) is relieved, and displacements and joint rotations are sought such as to establish an equilibrium for all forces and moments acting on each ridge prism (Fig. 2.7a). Here the external $\{R\}$ represent the external joint loading of the structure and the reactions due to the load acting between the joints, and the internal $\{\bar{S}\}$ depend on the deformations $\{\bar{\delta}\}$ (see relation (2.38)). The analysis can be generalized, when the displacements $\{r\}$ (Fig. 2.7b) of some joints are prescribed and the reactions $\{X\}$, acting at these joints, are sought.

Bearing in mind that the edge of one element is identical with the edge of another element, and that at the joint the continuity of the structure must be preserved, which means that the displacements and rotations of the contacting edges must be equal, the structure stiffness matrix $[K]$ can be arrived at by assembling the stiffness matrices $[k]$ of the elements.

The structure stiffness matrix makes it possible to write the following relationship between the ridge forces $\{R\}$ and the displacements $\{\bar{\delta}\}$

$$[K]\{\bar{\delta}\} = \{R\}. \tag{2.44}$$

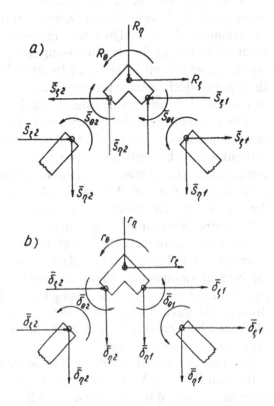

Fig. 2.7. Equilibrium and compatibility of a folded plate structure in a ridge.

With all the quantities (loading, internal forces and deformations) being expanded into Fourier series, in which the n-th term of one quantity is linked only with the n-th term of other quantities (that is, members of the series with different n values do not affect each other), the solution can be carried out for each harmonic separately. Relationship (2.44) written for the n-th term represents a system of linear algebraic equations, in the form

$$[K_n]\{\bar{\delta}_n\} = \{R_n\} \tag{2.45}$$

for the unknown amplitudes $\{\bar{\delta}_n\}$ of the n-th term of the series expressing the displacements and rotations of the joints in the fixed coordinate system, where $\{R_n\}$ are the amplitudes of the terms of the series expressing the loading of the joints.

The solution of system (2.45) is

$$\{\delta_n\} = [K_n]^{-1} \{R_n\}. \tag{2.46}$$

Usually the loading $\{R\}$ of the joints is known, and it remains to evaluate the displacements $\{\delta\}$. Cases occur, however, in which the displacements of some joints are prescribed in advance (for example, when any joint is supported, or when only half of the width of a symmetrical structure, loaded symmetrically or antisymmetrically, is analysed, with the corresponding component of displacement at the axis of the structure being zero). In this case, though, the corresponding components of the joint loadings $\{R\}$, representing the reactions at those joints which have the prescribed displacements, are unknown. The number of unknown quantities remains unchanged.

If the components of the displacements of some joints are zero, the conditions of equilibrium at the joints represented by the system of equations (2.45) may with advantage be written merely for the directions of the free displacement components. The number of unknown amplitudes in systems (2.45) is then reduced by the number of known displacement components. The reactions of these joints are then determined by means of forces or moments $\{\bar{S}\}$ at the edges of the elements which form these joints.

The analysis is almost complete when the required joint displacements $\{\delta\}$ are determined. It only remains to evaluate, using relation (2.38), the amplitudes of the forces and moments $\{\bar{S}_n\}$ at the edges of the individual elements, to determine the stresses of the elements according to (2.17) and (2.24), to add them to the stress for the state in which the elements to be clamped along the longitudinal edges have been considered and, finally, to sum the Fourier series in which all the quantities were assumed.

The ridge prisms, which have been assumed to match the individual elements at the joint (Fig. 2.1b), were tacitly supposed in the above to be massless, with no cross-sectional dimensions and no stiffness.

This assumption is fully justified in current cases when the plates of the actual folded plate structure intersect without any substantial and sudden strengthening at the joints (Fig. 2.2a). For more massive structures with actual ridge prisms (Fig. 2.8a), their effect should be taken into account. The plate elements are separated from the ridge prisms (Fig. 2.8b), and the stiffness matrix of the elements is set up so that the points of intersection of the element median lines are subjected to deformations with unit amplitudes. The elements, which in this case are narrower than the separations of the points of intersection, are assumed to be connected with these points by a system of rigid links (Fig. 2.8c shows the deformation pattern corresponding to conditions (2.10) and (2.12)).

Next, it is necessary to set up the stiffness matrix of the ridge prisms; this matrix relates the deformations of the prisms (longitudinal, vertical and horizontal displacements and rotations of the intersection points of the median lines) and the load acting on them. This can be effected by using the ordinary theory again, separately for each term of the Fourier series. It must be borne in mind that the intersection point of the element median lines is in general not identical with the centre of gravity, or with the shear centre of the prisms (Fig. 2.8b). Therefore, when assembling the stiffness matrix, a combination of the basic stress cases (such as tension accompanied by bending, etc.) should be considered.

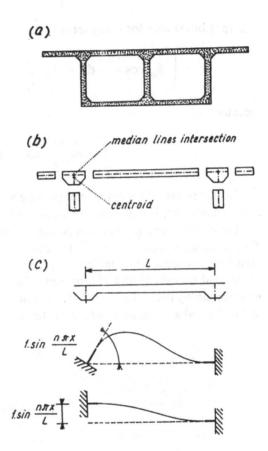

Fig. 2.8. (a) Cross-section of a structure with ridge prisms, (b) division of the structure into elements and edge prisms, (c) deformation of the element due to the displacements of longitudinal edges with unit amplitudes.

The stiffness matrices of the ridge prisms are to be taken into consideration when the stiffness matrix $[K]$ of the entire structure is assembled. The system of linear algebraic equations for the unknown deformations of the joints again expresses the conditions of equilibrium of the joints, but with regard to the stiffness of the ridge prisms.

The analysis (conducted in the manner described) makes it possible to load the structure by arbitrarily distributed vertical and horizontal loadings, at the joints and between them, as well as by a moment loading at the joints. The longitudinal loading, however, is subjected to the condition that its total on each element or joint must be zero, i.e., the longitudinal load must be self-balancing on each element and joint of the structure. This follows from the fact that the longitudinal loadings are expressed by series of the type

$$\sum_{n=1}^{\infty} \beta_{cn} \cos \frac{n\pi x}{L},$$

where it holds true for every term that

$$\int_0^L \beta_{cn} \cos \frac{n\pi x}{L} \, dx = \beta_{cn} \int_0^L \cos \frac{n\pi x}{L} \, dx = 0, \tag{2.47}$$

because

$$\int_0^L \cos \frac{n\pi x}{L} \, dx = 0. \tag{2.48}$$

For this reason the longitudinal loading which does not satisfy the aforementioned condition, but whose sum for the entire structure equals zero, should be split into two stages. (A typical instance of longitudinal loading whose sum total for the entire structure is equal to zero, but need not be so for each joint, is structure prestress.) In the first stage, the actual loading (Fig. 2.9a) is complemented by a shear flow of constant value, which en⎡ures that each joint is in equilibrium by itself (Fig. 2.9b); the second stage involves the oppositely acting shear flow which has been added in the first stage (Fig. 2.9c).

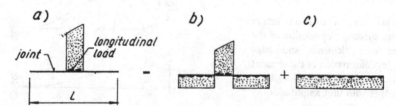

Fig. 2.9. Components of the longitudinal load acting at a joint.

The first stage (Fig. 2.9b) can be analysed in the manner described. A point of interest is that, while expressing the first loading stage in a series for $n \geq 1$, it is sufficient to consider the original loading (Fig. 2.9a), because the coefficients of the series related to the constant shear flow are equal to zero (compare relation (2.48)) and the expression of the loading according to Fig. 2.9a converges directly to the loading according to Fig. 2.9b.

The folded plate structure (Fig. 2.16a) is divided into an appropriate number of plate-shaped elements (Fig. 2.16b), and the action of the elastic foundation is represented by a set of continuously distributed elastic supports, acting at the joints. These joint reactions generally have four components, which correspond to interaction with the elastic foundation in the sense of four degrees of freedom (horizontal, vertical, rotational and longitudinal motions). The elastic properties of the joint supports are evaluated on the basis of the parameters of the actual elastic foundation supporting the halves of the elements adjacent to a joint.

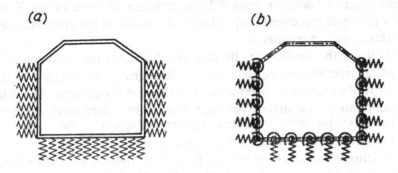

Fig. 2.16. (a) A folded plate structure on an elastic foundation, (b) folded plate idealization.

Due to the deformation of the folded plate structure, reactions in the continuous joint elastic supports appear, thus forming the additional joint loads $\{^eR\}$, whose magnitudes are proportional to the joint displacements (or to the joint rotation, respectively). Hence,

$$\{^eR\} = [^eK]\{\delta\}, \tag{2.69}$$

in which $[^eK]$ is a diagonal matrix of the individual components of the stiffness properties of the joint elastic supports, and $\{\delta\}$ are the joint displacements in the fixed coordinate system (Fig. 2.6). It follows that

$$[^eK] = \begin{bmatrix} ^ek_{\zeta 1}, & 0, & 0, & ..., & 0, & 0 \\ 0, & ^ek_{\eta 1}, & 0, & ..., & 0, & 0 \\ & & & & & 0 \\ 0, & 0, & 0, & ..., & 0, & ^ek_{\theta N} \end{bmatrix}, \tag{2.70}$$

where, for example, the term $^ek_{\eta 1}$ represents the stiffness of the elastic support under joint 1 against vertical displacement (see Fig. 2.16).

By expressing the loading and all functions describing the behaviour of the folded plate structure in the form of Fourier series, the solution for each term of the series can with great advantage be carried out separately and the partial results merely summed. The solution for each term is, within the sphere of the assumptions of the theory of elasticity, perfectly accurate. If N harmonics are used, an accurate solution of the behaviour of structure is obtained, but only for a loading expressed by N terms of the series and, therefore, differing from the actual loading of the structure. The difference between the solution obtained and the results which would be derived for $N \to \infty$, depends on the type of loading (different types of loading require different numbers of terms of the Fourier series for adequate representation) and on the character of the internal force or deformation which is compared.

The results of the solution, i.e., the functions describing the behaviour of the folded plate structure, can be classified in several groups with regard to their sensitivity to the number of terms of the series used in the analysis. The values of the deformations are already well approximated by a very small number of terms. To express the stress in the structure elements related to the   membrane effect, a higher number of terms in the series is needed

This conclusion is based on experience from the analysis of a structure loaded by a concentrated force. In practice this force, however, is always distributed over a finite area of the actual structure. In this way, the real situation is better expressed and, simultaneously, satisfactory results can be obtained with a smaller number of terms of the series.

2.2   Folded Plates with Deformable End Cross-Sections

Applications of the method presented in Section 2.1 are restricted to structures whose end cross-sections include transverse diaphragms infinitely rigid in their own plane and perfectly flexible out of this plane, preventing any distortion of the shapes of end cross-sections. This requirement, resulting from the application of harmonic analysis, represents a rather severe limitation and disadvantage of the folded plate theory.

In practice, cases are encountered where the requirement of diaphragms infinitely rigid in their own plane cannot be fulfilled. Deformable diaphragms are often used and, in some cases, the diaphragms are eliminated for various reasons (e.g. in double-level box girder bridges where free space must be maintained inside the cell, or as a production simplification in precast prestressed concrete members).

A modification of the folded plate theory, described in this Section, enables the effects of deformable end cross-sections to be predicted   from   simple calculations (Křistek (1983)).

The classical folded plate theory shown in Section 2.1 is a well established approach to the analysis of thin-walled prismatic structures with rigid end diaphragms.

The calculation, assuming end diaphragms rigid in their own planes, may be regarded as the first step of the analysis of folded plates with deformable end cross-sections.

In the second step of the analysis, the folded plate structure assumed has either no end diaphragms or deformable end diaphragms. The end cross-sections of the structure are loaded by forces which are transmitted by the non-existent rigid end diaphragms as obtained in the first step.

The resulting state is the sum of the states in both steps of the analysis.

The idea of the analysis in the second step is based on the well known fact that a distortional loading uniformly (Fig. 2.11a) or linearly (Fig. 2.11b) distributed over the length of a prismatic thin-walled girder with free ends and without any diaphragms, produces only the uniform distribution (Fig. 2.11c) or the linear variation (Fig. 2.11d) of distortional deformations. These particular states manifest themselves only by the distortion of cross-sectional shapes, and no longitudinal stresses arise.

Regarding these particular cases, the effects of distortional forces acting at the end cross-sections without diaphragms can be evaluated.

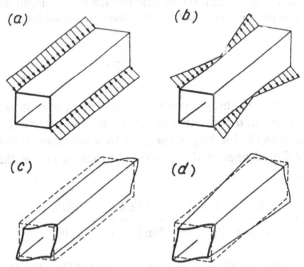

Fig. 2.11. A thin-walled girder without any diaphragm:
(a) uniformly distributed distortional loading,
(b) linearly distributed distortional loading,
(c), (d) corresponding deformation patterns.

2.3   Folded Plates with Diaphragms Restraining Free Cross-Sectional Warping

In practice, diaphragms are usually assumed to be rigid in their own plane and perfectly flexible out of this plane. This means that the diaphragm prevents deformation of the shape of the cross-section at which it is placed, but does not induce any bimoment effects. These assumptions are in accordance with the restrictions of the classical folded plate theory presented in Section 2.1, but often they are far from reality. The problem of end diaphragms which are deformable in their own planes was discussed in Section 2.2. The analysis of the structural performance of folded plates with thick or massive diaphragms, which exhibit a corresponding out-of-plane stiffness, is presented in this Section.

A diaphragm (and also a diaphragm with an opening, a frame cross-bracing or a truss) is a planar figure having its in-plane and out-of-plane stiffnesses. The diaphragm is connected to the folded plate structure, either at the joints only or along the whole perimeter of the cross-section. The diaphragm properties are described by the stiffness matrix, which relates the displacements of points on the edge of the diaphragm to the forces acting at those points.

Suppose the diaphragm thickness (in relation to its other proportions) permits us to consider it as a plate within the meaning of the theory of elasticity. Then, assuming a linear distribution of warping of the girder cross-section between two adjacent corners (Fig. 2.12a), the middle surface of the diaphragm will acquire the form of a hyperbolic paraboloid described by the equation

$$u(\eta, \zeta) = -4\bar{u}\,\frac{\eta\zeta}{lh}, \tag{2.51}$$

in which l, h are the diaphragm dimensions, and η, ζ the local coordinates (Fig. 2.12a). This pattern of deformation of the diaphragm corresponds, according to the plate theory, to the loading of the plate by longitudinal forces R (shown in Fig. 2.12a by dotted lines in the direction of their action on the diaphragm), having a value

$$R = \left| \frac{Et^3}{6(1 + \bar{v})}\,\frac{\partial^2 u}{\partial\eta\,\partial\zeta} \right| = \frac{2E t^3\bar{u}}{3lh(1 + \bar{v})} = \frac{4\bar{G}t^3\bar{u}}{3lh} = k\bar{u}, \tag{2.52}$$

where $\bar{G}$ is the shear elasticity modulus of the diaphragm material and k is the stiffness of the diaphragm for a plate bending effect.

Due to the warping of the box girder cross-section, the hyperbolic paraboloid deformation pattern of a plate according to Eq. (2.51) induces either reactions R acting in the plate corners (Fig. 2.12a, formula (2.52)) or specific torsional reaction moments of an intensity $Et^3/12(1 + \bar{v}) \cdot \partial^2 u/\partial\eta\,\partial\zeta$ acting along the contour of the plate (according to the classical plate theory both systems of reactions are equivalent). In this state of stress, only torsional moments (equal in value to the torsional reaction moments acting on the plate edges) appear in the plate; these moments induce only shear stresses $\tau_{\eta\zeta}$ parallel to the middle plane of the plate. The distribution of these stresses along the plate thickness is, in this particular simple case, linear. This is true not only if the classical theory

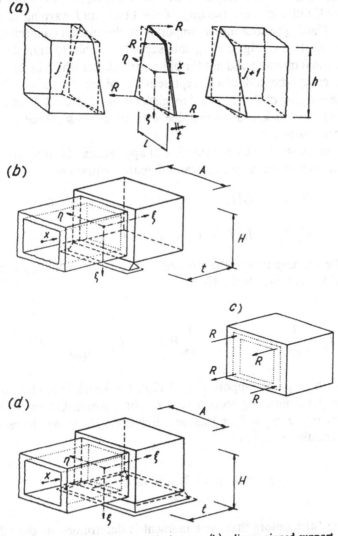

Fig. 2.12. (a) Slab action of a disphragm, (b) a linear pinned support of end blocks, (c) interaction between the box girder and an end block, (d) an end block supported along its entire periphery.

is employed, but also if use is made of more precise theories (such as the component theory, Reissner's or Hencky's theories, etc.); consequently, one has $\tau_{\zeta\eta} = \bar{E}/(1 + \bar{v}) \cdot (\partial^2 u/\partial\eta \, \partial\zeta) \cdot \bar{x}$, where $\bar{x}$ is the coordinate measured perpendicular to the middle plane of the plate. Therefore, the relations derived are also suitable for diaphragms with somewhat greater thickness, which for another kind of plate-type loading cannot be considered as plates.

When the thickness of the diaphragm is increased, the end diaphragms acquire the character of massive blocks (schematically shown in Fig. 2.12b). It will be assumed in the analysis that the cross-sectional shapes of this block are not deformed, but that only twisting of the block and warping of its cross-section take place. The structure is supported in the section connecting the box girder with the block (Fig. 2.12b); this support is able to transfer only the support torsional moment and not the bimoment effects which in this case must be taken over by the end block, representing an elastic fixing. This fixing restrains the warping of the extreme box girder cross-section. It is assumed that the length of the block t is comparable with its cross-sectional proportions height H and width A.

If the end block is taken to be a member of approximately solid cross-section, its behaviour may be described by two differential equations

$$\bar{E}I_\omega f'' - \bar{G}[(I_\eta + I_\zeta) f - (I_\eta - I_\zeta) \varphi'] = 0,$$

$$\bar{G}[(I_\eta - I_\zeta) f' - (I_\eta + I_\zeta) \varphi''] = m_k, \tag{2.53}$$

where f is the warping rate of the cross-section, φ is the twisting angle and m_k the torsional load acting on the block.

Then

$$I_\eta = \frac{1}{12} AH^3; \qquad I_\zeta = \frac{1}{12} HA^3; \qquad I_\omega = \frac{1}{144} A^3 H^3.$$

In view of the type of support (Fig. 2.12b), the block is loaded only by a bimoment at the connecting section with the box girder. Otherwise it is not loaded and therefore $m_k \equiv 0$ in this case. Relations (2.53) may be rearranged into one differential equation

$$\bar{E}I_\omega f''' - \bar{G} \frac{4I_\eta I_\zeta}{I_\eta + I_\zeta} f' = 0. \tag{2.54}$$

Considering furthermore that the bimoment is determined by the relation

$$B = \bar{E}I_\omega f', \tag{2.55}$$

Eq. (2.54) may be further adjusted to the form

$$\bar{B}'' - b^2\bar{B} = 0,$$ (2.56)

where

$$b = \sqrt{\frac{\bar{G}}{E} \frac{4I_\eta I_\xi}{I_\omega(I_\eta + I_\zeta)}} = \sqrt{\frac{48\bar{G}}{(A^2 + H^2)E}}.$$ (2.57)

The solution of this equation, using a new longitudinal coordinate ξ with the origin at the contact section between the end block and the box girder, is

$$\bar{B}(\xi) = C_1 \cosh b\xi + C_2 \sinh b\xi.$$ (2.58)

If forces R act on the block at contact points with the corners of the box girder median line according to Fig. 2.12c (that is at points with coordinates $\xi = 0$, $\eta = \pm a/2$, $\zeta = \pm h/2$), they induce a bimoment with the value of

$$\bar{B}(0) = 4R \frac{a}{2} \frac{h}{2} = Rah,$$ (2.59)

which forms one boundary condition for the determination of constants in the general solution (2.58). The second condition, at the free end of the block $\xi = t$ where no bimoment is acting, is

$$\bar{B}(t) = 0.$$ (2.60)

The bimoment then develops along the block length according to the relation

$$\bar{B}(\xi) = Rah(\cosh b\xi - \cotgh bt \sinh b\xi).$$ (2.61)

Longitudinal displacements of the block points – in this case, warping – may be described by the relation

$$u(\xi, \eta, \zeta) = f(\xi) \omega(\eta, \zeta) = f(\xi) \eta\zeta.$$ (2.62)

By substituting Eq. (2.55) into Eq. (2.62), using Eq. (2.61) and integrating, while noting that on the block axis ξ, 0, 0, no longitudinal displacements take place, the following equation is obtained

$$u(\xi, \eta, \zeta) = \frac{Rah}{EI_\omega b} (\sinh b\xi - \cotgh bt \cosh b\xi)\eta\zeta.$$ (2.63)

In the place of contact with the upper point of the left box girder web (that is at a point with coordinates $\xi = 0$ $\eta = a/2$ $\zeta = -h/2$), this results in

$$u\left(0, \frac{a}{2}, -\frac{h}{2}\right) = \frac{Ra^2h^2}{4\bar{E}I_\omega b} \cotgh bt . \tag{2.64}$$

This displacement must equal the longitudinal displacement $\bar{u}$ of the corner point of the end box girder cross-section. From this the following relation is derived

$$R = \frac{4\bar{E}I_\omega b\bar{u}}{a^2h^2} \tgh bt = k\bar{u}, \tag{2.65}$$

which is an analogue of Eq. (2.52), where k again means the stiffness of the block.

If the entire length of the block is supported (Fig. 2.12d), the solution may again be found according to Eq. (2.53), where in this case $\varphi = 0$ is to be taken, because this type of support prevents twisting of the block. From the second equation (2.53), the intensity of a torsional load m_k may be expressed; this load is needed to prevent twisting of the block and it must be transferred by the support reactions. It may be written as follows:

$$m_k = \bar{G}(I_\eta - I_\zeta) f . \tag{2.66}$$

The first equation then takes the simple form:

$$\bar{E}I_\omega f'' - \bar{G}(I_\eta + I_\zeta) f = 0 , \tag{2.67}$$

which, after differentiation and application of relationship (2.55), assumes a form similar to Eq. (2.56). Also solution (2.58) and all the other relationships retain their validity, but the quantity b must be substituted by another quantity

$$\bar{b} = \sqrt{\frac{\bar{G}}{E} \frac{I_\eta + I_\zeta}{I_\omega}} = \sqrt{\frac{12\bar{G}(A^2 + H^2)}{\bar{E}A^2H^2}} . \tag{2.68}$$

If the block has a square cross-section (that is $H = A$), then both types of support described are equivalent, because a block with a square cross-section under a bimoment loading of the end cross-section does not tend to twist. This also results from relation (2.66), from which $m_k = 0$ is obtained and

$$b = \bar{b} = \sqrt{\frac{24\bar{G}}{\bar{E}A^2}} .$$

If the relationship between the forces and the displacements at the diaphragm corners is known, the interaction between the box girder and the diaphragm can be solved. For this purpose, for example, the force method of analysis may be used, in which the bonds between the box girder and the bracings are stated, and the redundant parameters acting in these bonds which maintain the continuity of the structure are sought.

2.4 Folded Plate Structures with Various End Conditions

The classical folded plate theory presented in Section 2.1 is restricted to simply supported structures, provided with end diaphragms which are infinitely rigid in their own planes and perfectly flexible out of those planes. The methods presented in Sections 2.2 and 2.3 allow for analysis of folded plates with deformable end diaphragms (or without any diaphragms) and diaphragms with an out-of-plane stiffness. However, there are some other arrangements of end conditions which can easily be solved by establishing appropriate substitute structural models.

For example, a folded plate structure with overhanging ends (Fig. 2.13a) can be analysed by taking its total length as the span of a substitute, simply supported structure loaded both by the given external loading and by the known statically determinate reactions, Fig. 2.13b. Thus a set of self-equilibrating forces is formed and, consequently, no reactions appear at the end cross-sections of the substitute structure. The states of stress of both the structure with overhanging ends and the substitute simply supported structure are identical; the displacements thus obtained (Fig. 2.13d), however, are to be corrected by rigid body motion, realizing that no deflections originate in the actual supports, Fig. 2.13c.

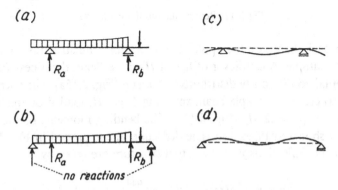

Fig. 2.13. Representation of a structure with overhanging ends:
(a) a structure with overhanging ends, (b) the substitute structure,
(c) deflections of the original structure, (d) deflections
of the substitute structure.

Similarly, a folded plate structure formed as a cantilever, Fig. 2.14a, (fixed at one end and free at the other end, with a diaphragm at the free end) may be solved as a simply supported folded plate structure spanning double the length of the cantilever (the original cantilever and its mirror image about the clamped end, Fig. 2.14b) provided with a diaphragm rigid in its own plane at mid-span. This substitute structure is loaded by the given external load acting on the two cantilevers formed in this way and, at the mid-span of the simply supported structure, where the diaphragm is situated, by force factors acting in the diaphragm plane (e.g. three forces, or two forces and one moment) whose combined effect double the value of the resultant of the reversedly taken given external loading of the cantilever. Since these additional factors balance all the applied loads, the resulting reactions of the substitute structure are then equal to zero (which coresponds to the free end of the cantilever) and the state of stress of the half span corresponds to the actual cantilever structure analysed. The deformations of the two structures differ only in that the maximum deflection of the original cantilever occurs at the end, as in Fig. 2.14c, whereas the ends of the substitute structure are supported (Fig. 2.14d).

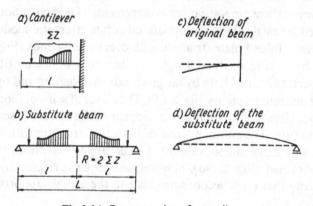

Fig. 2.14. Representation of a cantilever.

This approach, based on the application of a substitute structure, can be shown by a simple example. A cantilever of height H with a flexural stiffness EI is loaded by a horizontal continuously distributed loading q (Fig. 2.15a). The corresponding substitute structure is a simple beam spanning $L = 2H$, loaded by the load q and the force $R = qL = 2qH$ (Fig. 2.15b). The bending moments of the substitute structure are shown in Fig. 2.15c. The deflection line is plotted in Fig. 2.15d. It is evident that the deflections y of the actual structure are given by

$$y(x) = \bar{y}(H) - \bar{y}(x) \cong \frac{qH^4}{8EI} - \bar{y}(x) ,$$

where $\bar{y}(x)$ is the deflection of the substitute structure.

This idea is applied to the analysis of tall buildings as shown in Section 2.7.

The introduction to this Chapter dealing with the analysis of folded plate structures included the basic requirement of a constant cross-section of the structure over its entire length. In the special case of a continuous folded plate structure provided at all its supports with diaphragms perfectly rigid in its own plane and perfectly flexible perpendicularly to this plane (and having no other diaphragms), an approximate procedure may be applied. In this procedure folded plate structures, simply supported as individual spans and themselves fulfilling the basic assumptions of the folded plate theory, are chosen as the primary system (compare the analysis of a continuous girder by the three-moment equations). The redundant quantities in this case are represented by a set of longitudinal forces at selected points of the support cross-sections (representing support bending moments and support bimoments in the ordinary theory). This approach has the advantage of the flexibility matrices being of a band form; moreover, this selection of the primary structure makes it possible to have different thicknesses of walls in the individual spans, that is, the folded plate structure analysed may change its cross-section at the supports.

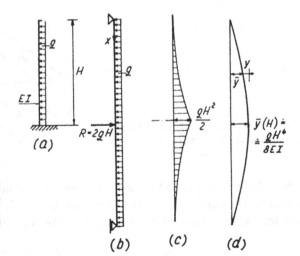

Fig. 2.15. Analysis of a cantilever beam: (a) the actual structure, (b) the substitute structure and its loading, (c) the bending moment diagram, (d) deflections.

2.5 Folded Plate Structures on an Elastic Foundation

Cases are encountered in practice where prismatic thin-walled, cellular or similar structures (tunnels, foundation structures under high-rise building, pipelines, subways, etc.), are placed on a foundation which may be regarded as elastic, i.e. in which induced reactions are proportional to deflections. The folded plate theory again provides a simple tool for structural analysis of these cases.

In the second loading stage (Fig. 2.9c) the structure is loaded by the shear flows $\bar{q}$ of constant values in the longitudinal direction, which act at the ridges of the folded plate structure (Fig. 2.10a). Because of the manner of support of the transverse edges of the elements on the end diaphragms of the structure, which permits longitudinal displacements of the end points of the elements to take place (perpendicular to the diaphragm plane) but controls all action in the plane of the diaphragms, a simple pure shear stress is produced in.the elements due to this loading.

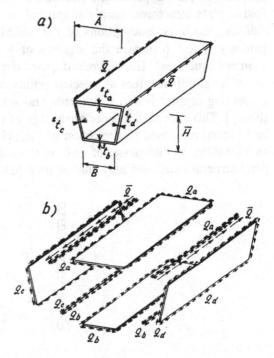

Fig. 2.10. Loading of the edges of a folded plate structure by constant longitudinal flows.

Each element is stressed by a shear flow of constant value for the respective element. The equilibrium of the ridge prisms (Fig. 2.10b) results, for example, in the following relations for the structure shown in Fig. 2.10a

$$q_a - q_c = \bar{q}, \qquad q_a - q_d = \bar{q},$$

$$q_c - q_b = 0, \qquad q_b - q_d = 0. \qquad (2.49)$$

Only three of these relations are independent, and they must therefore be complemented by another relation, which follows from the condition of continuity

$$\frac{2(1 + {}^s v)}{{}^s E} \left[\frac{q_a}{{}^s t_a} \bar{A} + \left(\frac{q_c}{{}^s t_c} + \frac{q_d}{{}^s t_d} \right) \sqrt{\left(\frac{\bar{A} - \bar{B}}{2} \right)^2 + \bar{H}^2} + \frac{q_b}{{}^s t_b} \bar{B} \right] = 0. \quad (2.50)$$

Since all the joint displacements are expressed in the form of Fourier series, and since the elements of matrix $[^eK]$ are constants, Eq. (2.69) may be written for the n-th harmonic in the form

$$\{^eR_n\} = [^eK_n]\{\bar{\delta}_n\} = [^eK]\{\bar{\delta}_n\}. \tag{2.71}$$

Analogously to Eq. (2.45), the equilibrium of each elastically supported ridge prism yields the relation

$$[K_n]\{\bar{\delta}_n\} = \{R_n\} - \{^eR_n\} = \{R_n\} - [^eK]\{\bar{\delta}_n\}, \tag{2.72}$$

in which $[K_n]$ is the structure stiffness matrix corresponding to the n-th harmonic, and $\{\bar{\delta}_n\}$ are the unknown amplitudes of the n-th term of the series expressing the displacements and rotations of the joints, see Section 2.1. Equation (2.82) can be written as

$$([K_n] + [^eK])\{\bar{\delta}_n\} = \{R_n\}, \tag{2.73}$$

or in the form

$$[\bar{K}_n]\{\bar{\delta}_n\} = \{R_n\}, \tag{2.74}$$

which is formally identical with Eq. (2.45).
The matrix

$$[\bar{K}_n] = [K_n] + [^eK] \tag{2.75}$$

represents the stiffness matrix of the folded plate structure on an elastic foundation. It can be simply obtained by adding the elastic support characteristics forming the diagonal of matrix $[^eK]$ to the main diagonal elements of the structure stiffness matrix $[K_n]$.

Thus it is clear that the analysis of folded plates on an elastic foundation can again be carried out by introducing only minor adjustments (replacement of the original stiffness matrices $[K_n]$ by matrices $[\bar{K}_n]$, Eq. (2.75)) into existing folded plate computer programs, and that the method presented retains all the advantages of the classical folded plate theory.

The harmonic analysis in the form presented requires that simple supports be placed under the end diaphragms at both ends of the structure (Fig. 2.17a). In practice, however, arrangements without end supports (Fig. 2.17b) or, in some cases, also without end diaphragms (Fig. 2.17c) are more frequent.

An idea similar to that applied in Section 2.2 for the analysis of folded plates with deformable end cross-sections can be used. Considering, for example, the case shown in Fig. 2.17b, it is possible to supplement the given actual loading by such a set of uniformly or linearly distributed joint loadings (or their combinations of a trapezoidal form) such that no reactions appear at the end supports (Fig. 2.17d) – despite the fact that the structure is simply supported at the ends.

The set of trapezoidally distributed loadings which was added to the actual loads in order to bring about the special state without any reactions at the simply supported ends (Fig. 2.17d) must be applied in the opposite direction with respect to the structure in its actual form, i.e. without end supports (Fig. 2.17e). These trapezoidal loadings produce only deflections of linear variation, without inducing any longitudinal or shear stress in the structure.

The case without supports and end diaphragms (Fig. 2.17c) can also be analysed in a similar way, i.e. by application of the idea presented in Section 2.2.

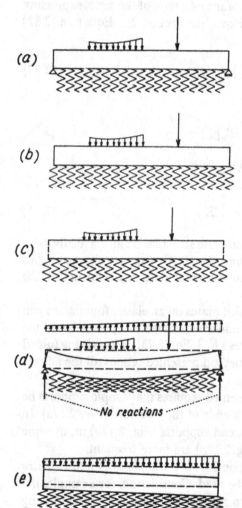

Fig. 2.17. (a) A simply supported folded plate structure (with support diaphragms) on an elastic foundation, (b) a structure without end supports, (c) a structure without end supports and without end diaphragms, (d) a simply supported structure with supplementary loading, (e) the action of trapezoidally distributed loading on the structure without end supports.

2.6   Statically Indeterminate Continuous Folded Plate Structures with Interior Diaphragms

Statically more complicated folded plate structures (continuous, clamped structures, frames, girders with interior right or skew diaphragms having plate stiffness, or with flexible support frame bents, etc.) can be analysed by means of the force method. This method consists of removing all redundant constraining effects (supports, bracings, etc.) so that a simply supported folded plate structure is obtained. Its analysis was described in Section 2.1 (or, for the case with deformable end diaphragms, in Section 2.2), and this structure will now represent the primary system for the solution by the force method. The removed constraining effects are replaced by the redundants $\{Y\}$, whose magnitudes are to be determined from the deformation conditions. The primary system (the simply supported folded plate structure) being analysed by means of a series, the reality may be better expressed by distributing the redundants over a given length (such as the width of the bearing). The redundants are to be chosen so as not to induce any longitudinal force in the primary structure as a whole. The method is demonstrated on two most frequent cases of statically indeterminate structures.

2.6.1   Folded plate structure with frame bents

The structure is continuous with interior frame bents for supports (Fig. 2.18a). These frame bents are assumed to be perfectly flexible in the direction perpendicular to their own planes. If the rigidity in the longitudinal direction, in the case of the structure analysed, cannot be neglected, it may be simulated approximately by doubling the frame bents, each of them having half rigidity in the transverse direction, with the flexural rigidity in the longitudinal direction being substituted by the axial rigidity of the columns (rigidity in tension or compression), Fig. 2.18b.

To form the primary structure, the frame bents are separated from the folded plate structure in the support cross-sections as shown in Fig. 2.18c; the actual shape of the frame bents is idealized by planar frames corresponding to the actual arrangement. For almost rigid links, for example, which connect the frame with the nodal points of the folded plate structure (Fig. 2.18d) and which are actually situated within the support bracing, a very high value of the modulus of elasticity can be introduced.

The interaction forces (redundants) $\{Y\}$ are represented by a set of three forces consisting of vertical, horizontal and rotational components in the plane of the transverse cross-section.

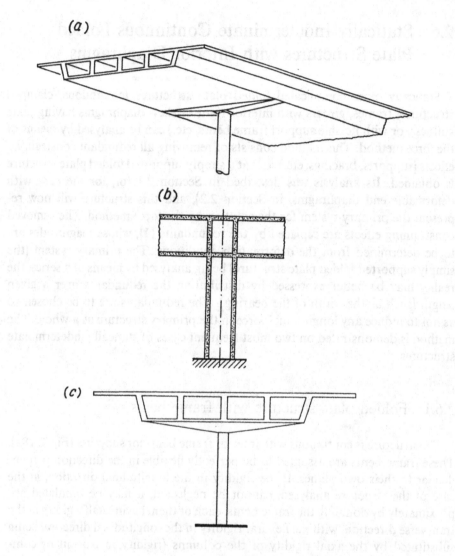

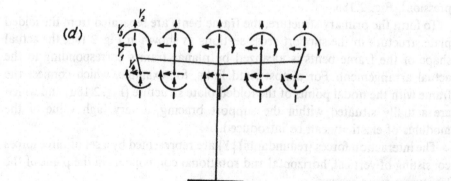

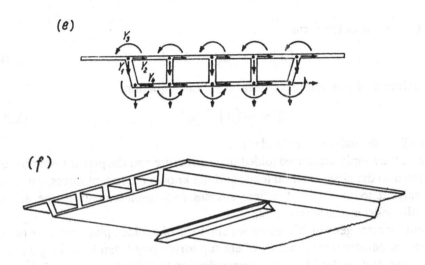

Fig. 2.18 (a) A frame bent supporting a folded plate structure, (b) an idealized frame bent having a rigidity in the longitudinal direction, (c) the support cross-section of the structure, (d) an idealized frame bent and the action of the redundants, (e) the effect of the redundants on the folded plate structure, (f) an interior pinned support of a continuous folded plate structure.

The primary, simply supported single-span structure is analysed first for the given external loads acting alone (with all redundants removed). A displacement vector $\{d_0\}$ is found for this case, which defines the displacements of the points where the redundants are to act. By successively applying unit redundant loads to the system (Fig. 2.18e) and solving it for displacements, the individual columns of the structure flexibility matrix $[\Delta_1]$ are established.

Each of the planar frame bents is analysed by the direct stiffness method. The total structure stiffness for the frame bent is found, then static condensation is carried out to eliminate the degrees of freedom which do not correspond to the redundant forces. The flexibility matrix of the frame bents $[\Delta_2]$ corresponding to the unit redundant forces is found by inverting the stiffness matrices

Geometric compatibility at the interaction points requires the fulfilment of the conditions

$$[\Delta_1]\{Y\} + [\Delta_2]\{Y\} + \{d_0\} = 0,\qquad(2.76)$$

which, with the notation

$$[\Delta] = [\Delta_1] + [\Delta_2],\qquad(2.77)$$

can be written in the form

$$[\varDelta]\{Y\} + \{d_0\} = 0.$$ (2.78)

The solution of this system of linear algebraic equations

$$\{Y\} = -[\varDelta]^{-1}\{d_0\}$$ (2.79)

gives all of the values of redundants.

Thus, the simply supported folded plate structure and the planar frame bents, subjected to the given external loads and the known redundant forces, can now be analysed to determine the final stresses and displacements in the actual statically indeterminate structure.

Another arrangement of the interior supports of a folded plate structure often appears in construction practice, where supports carry a diaphragm rigid in its own plane and perfectly flexible perpendicular to this plane (Fig. 2.18f); the structure has the character of a continuous beam. All points of the support cross-section are prevented, by the rigid diaphragm, from being displaced in the cross-section plane; therefore, the interaction points of the folded plate structure also cannot be displaced horizontally and vertically, nor can they rotate, at the points at which the redundants Y (Fig. 2.18e) act. The problem may then be solved in the same way as that of the frame bents, except that $[\varDelta_2] = 0$ is to be introduced in Eq. (2.76).

The connection of the support diaphragms with only some interaction points of the folded plate structure (if need be, only in the sense of some reaction components) makes it easily possible to simulate various other arrangements of the supports encountered in practice.

2.6.2   Folded plate structure with interior (unsupported) diaphragms deformable in their own planes

The intermediate diaphragms (Fig. 2.19a) are assumed to be perfectly flexible in the directions perpendicular to their own plane, but unlike the diaphragms at the end cross-sections, they are deformable in their own plane. Since these diaphragms are not externally supported they can undergo, when subjected to interaction forces, three degrees of rigid body motion in their own plane in addition to the deformation of the diaphragms themselves. Therefore, the interaction forces must be in self-equilibrium.

For this reason, the analysis must be conducted in a manner differing somewhat from that of the frame bents. Not all of the restraints between the diaphragm and the folded plate structure are released; three restraints, representing the statically determinate support of the diaphragm on the folded plate

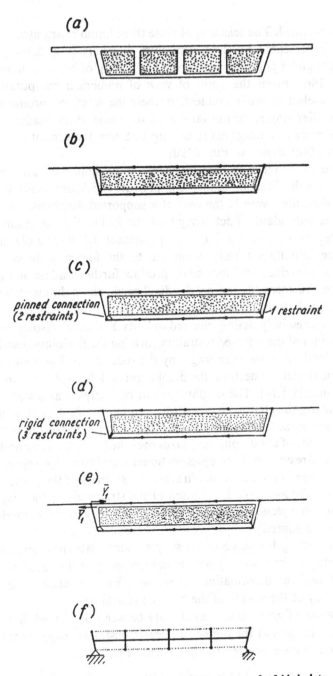

Fig. 2.19. (a) An interior unsupported diaphragm of a folded plate structure, (b), (c), (d) statically determinate supporting of a diaphragm, (e) the action of the redundants, (f) an example of idealization of a diaphragm by an equivalent frame.

structure, are retained. The selection of these three initial restraints is arbitrary; they may be represented by three pendulum members (Fig. 2.19b), or by a pinned support and a pendulum member (Fig. 2.19c), or by some fixing at one point (Fig. 2.19d). From the point of view of numerical computation, it is appropriate to select an initial connection where the reactions, produced by the loading of the diaphragm, do not far exceed the value of the loading in order. It is therefore more advantageous to use, for instance, the variant according to Fig. 2.19c than that shown in Fig. 2.19d.

At the points of the released restraints between the diaphragm and the folded plate structure (in the direction of their components which, as already stated, are less by three than they were in the case of a supported diaphragm or a frame bent), there the redundants $\bar{Y}$ act. They load the folded plate structure as well as the diaphragms (e.g., in Fig. 2.19e the redundant $\bar{Y}_1$). As the diaphragm is assumed to be initially statically connected to the folded plate system, the diaphragm reactions due to the redundant produce further loading on the folded plate structure. All forces acting on the diaphragm, the redundant $\bar{Y}$ and the reactions produced by this redundant, are in a state of self-balance.

Due to the successively acting unit redundants $\bar{Y}$, relative displacements $\bar{d}_{ki}$ occur at the points of the released restraints, forming the flexibility matrix $[\bar{\Delta}_1]$.

Due to the loading of the diaphragm by the redundants $\bar{Y}$ as well as by the reactions of the initial connection, the diaphragm is deformed, contributing to the flexibility matrix $[\Delta_1]$. The diaphragm can be analysed as a wall. Longer diaphragms having the character of a girder can be idealized as an equivalent frame, for example, statically determinate (Fig. 2.19f), where the horizontal member (elastic axis of the diaphragm) takes over the bending, shear and tensile or compressive stresses, and its cross-sectional characteristics represent the respective stiffnesses of the entire diaphragm. The stiffness of the perpendicular links depends on the assumed distribution of the strains along the diaphragm height (assuming that plane diaphragm cross-sections remain plane, totally rigid links shoud be introduced).

The external loading deforms only the simple folded plate structure, while the diaphragm, owing to its statically determinate support, is displaced only as a rigid body without any deformations of its own. Relative displacements $\{\bar{d}_0\}$ appear in this way at the points of the released restraints.

The preservation of geometric compatibility between the folded plate structure and the diaphragms at the points of released restraints requires the fulfilment of the conditions expressed by the system of equations

$$[\bar{\Delta}_1]\{\bar{Y}_1\} + \{\bar{d}_0\} = 0 \,. \tag{2.80}$$

It often happens that the structure is provided with both frame bents and interior unsupported diaphragms. It is, therefore, advantageous to maintain the same system of basic redundants throughout the analysis. For this reason, the procedure concerning the interior diaphragms can be the same at the beginning as the procedure concerning the frame bents. The primary folded plate structure, which excludes the diaphragm, is analysed under the external loading. The joint displacements $\{d_0\}$ at the location of the diaphragm are calculated, selecting the same system of redundants $\{Y\}$ as for the solution of the structure with frame bents. Then (again in the same way as when analysing the structure with frame bents), the flexibility matrix $[\varDelta_1]$ of the primary folded plate structure is formed. The displacements of the points of action of the redundants on the primary folded plate structure, produced by the total effect of all redundants, are described by the relation

$$\{d_{y,\text{fol}}\} = [\varDelta_1]\{Y\}. \tag{2.81}$$

Now, a new set of redundants $\{\bar{Y}\}$ is defined so that each redundant, while acting on the diaphragm, will be in self-equilibrium. Each of these new redundants $\bar{Y}$ has four components: the redundant Y and the diaphragm reactions which occur when the diaphragm is initially connected to the folded plate structure (Fig. 2.19b, c, d, e). The number of the redundants $\bar{Y}$ is therefore equal to the number of the redundants Y and is for each diaphragm less by three than the number of the redundants Y. The relation between the original redundants Y and the redundants $\bar{Y}$ can be expressed as

$$\{Y\} = [\psi]\{\bar{Y}\}, \tag{2.82}$$

where $[\psi]$ is a force transformation matrix, whose number of rows differs by three from the number of columns for each movable diaphragm.

The relative displacements $\bar{d}$ of the diaphragm and of the folded plate structure with the assumed initial connection of the diaphragm depend on the absolute displacements d of the points of the folded plate structure. The dependence is given by the relation

$$\{\bar{d}\} = [\psi]^T\{d\}. \tag{2.83}$$

Hence,

$$\{\bar{d}_0\} = [\psi]^T\{d_0\}, \tag{2.84}$$

$$\{\bar{d}_{Y,\text{fol}}\} = [\psi]^T\{d_{Y,\text{fol}}\}. \tag{2.85}$$

By introducing relations (2.76) and (2.82) into (2.85), the following equation is obtained

$$\{\bar{d}_{Y,\text{fol}}\} = [\psi]^T[\varDelta_1][\psi]\{\bar{Y}\}, \tag{2.86}$$

which, with the notation

$$[\psi]^T [\Delta_1] [\psi] = [\bar{\Delta}_1],\qquad(2.87)$$

can be written in the form

$$\{\bar{d}_{Y,\text{fol}}\} = [\bar{\Delta}_1]\{\bar{Y}\},\qquad(2.88)$$

where $[\bar{\Delta}_1]$ is the modified flexibility matrix of the simple folded plate structure (excluding the contribution due to the deformation of the diaphragm).

By analysing the sole diaphragm, initially connected to the folded plate structure and successively loaded by the unit values of the redundants, the displacements of the interacting points of the redundants are obtained. These displacements form the flexibility matrix $[\bar{\Delta}_2]$ of the diaphragm, hence the total effect of all redundants $\bar{Y}$ produces a displacement of the interaction points of the redundants

$$\{\bar{d}_{Y,zt}\} = [\bar{\Delta}_2]\{\bar{Y}\}.\qquad(2.89)$$

The geometrical compatibility between the folded plate structure and the diaphragm requires fulfilment of the conditions

$$\{\bar{d}_{Y,\text{fol}}\} + \{\bar{d}_{Y,zt}\} + \{\bar{d}_0\} = 0,\qquad(2.90)$$

which, after the introduction of relations (2.88) and (2.89) and the notation

$$[\bar{\Delta}_1] + [\bar{\Delta}_2] = [\bar{\Delta}],\qquad(2.91)$$

can be written in the form

$$[\bar{\Delta}]\{\bar{Y}\} + \{\bar{d}_0\} = 0,\qquad(2.92)$$

in which relation (2.82) can be introduced for the sake of uniformity of the system of redundants.

The final solution may be obtained by separately analysing the simply supported folded plate structure, the planar frame bents and the diaphragms being subjected to known external loading and redundants.

It should be noted, from the point of view of the numerical computation, that the flexibility matrices $[\Delta]$ should be established with a very high degree of accuracy; they may be so ill-conditioned that small changes of the individual coefficients may alter the magnitude of the redundants completely. If follows from this that either a large number of Fourier terms has to be used in the analysis, or fewer coupled sets of redundant force patterns should be selected in order to reduce the magnitude of the off-diagonal flexibility coefficients. Restraints whose corresponding redundants act against themselves on the stiff parts of the structure (for example, vertical restraints of both ends of the same web) should be avoided.

3. Finite element method

This method has been brought to a considerable perfection and has become a nearly universal method for the solution of the problems of mechanics in recent years. In its general form the method belongs in the group of direct variation methods and its principles are well known and systematically described.

The main advantage of the method is its universality: it lends itself to the solution of thin-walled structures of arbitrary shape, support conditions and load, as well as to the solution of their interaction with integrated one-, two- and three-dimensional elements (bracing ribs, frames, massive blocks, foundations co-operating with subsoil, etc.). For the analysis of thin-walled structures the direct-stiffness variant of the method is usually used on account of its numerous advantages; the primary unknown quantities in this case are the generalized displacements.

The computation procedure does not substantially differ from the analysis of other structural systems by this method and it can be summarized in several steps:

1. The structure is divided longitudinally into a number of segments which are further divided into individual (e. g. quadrilateral) elements (Fig. 3.1a). The numbering of the nodal points of the elements within the cross section should be arranged so as to have the smallest possible differences of the numbers which mark the nodal points; in this way a minimum bandwidth of the system of equations is obtained.

2. Stiffness matrices of all the elements are set up and transformed to a global common co-ordinate system.

3. According to the principles of the direct-stiffness method, by assembling the stiffness matrices of the elements the stiffness matrix of the structure is formed.

4. The right-hand side of the equations (the load vector), which represents the loading of the structure, is set up; the loading distributed over the area of the element is substituted by equivalent nodal loads. If there are several loading cases, a load matrix is established.

5. Boundary conditions of the problem are applied.

6. The system of linear algebraic equations is solved, best by a method which takes advantage of the banded nature of the stiffness matrix of the structure and in this way substantially reduces the computing time. Thus, the displacements of the nodal points are evaluated (i. e., the displacement vector).

7. By multiplying the stiffness matrix by the displacement vector, the reactions are obtained. The reactions at the points whose displacements were not prescribed have the character of residual loads which are a measure of the accuracy of the solution.

8. Reverse transformation of the nodal displacements is carried out from the global to the element co-ordinate system; the stiffness matrix is multiplied by the displacements and the stress of the element is established.

The quality of the analysis depends on the choice of the element model. Considering that a complicated spatial structure is analyzed and that its general behaviour (such as the bending of the structure as a whole), as well as the stress due to the spatial effect of the structure, and or to some extent also the effect of some irregularities, should be rendered, a relatively large number of elements is unavoidable in the analysis.

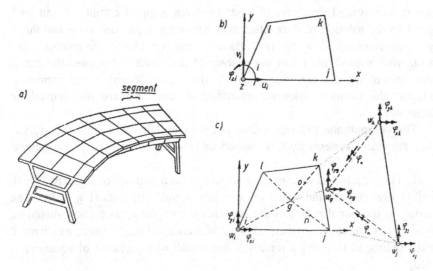

Fig.3.1 *a*. Division of a box structure into finite elements, *b*. quadrilateral plane stress element with twelve degrees of freedom, *c*. quadrilateral plate bending element with twelve degrees of freedom

The elements into which the structure had been divided, are considered plane; an element of a curved web may be also replaced by a plane projection into a plane tangential to the middle surface of the web, because in box girders the radii of curvature of the webs are incomparably larger than the dimensions of the element. Assuming a small-deflection linear theory, the membrane and plate bending behaviours of a plane element are mutually uncoupled.

Various methods may be used for the approximation of the behaviour of the element. One of them at least will be demonstrated as an example; the model is convenient as it renders the girder behaviour satisfactorily and has six degrees of freedom (three translational and three rotational) at each nodal point; this facilitates the attachment of three-dimensional frames to the investigated structure, as the member nodes of these frames have also six degrees of freedom.

The complete stiffness matrix of the element for membrane and plate effects is of the size 24 by 24.

Using the finite-element method it is possible to analyze almost any structure with sufficient accuracy.

4. Finite-strip method

This method is basically a transition between the folded-plate and the finite-element methods; it combines some of their advantages and presents an efficient way to the solution of many types of structures. The method is well established.

All structures satisfying the assumptions of the folded-plate theory (constant cross section, braced end cross sections, etc.) can be analyzed by this method (Fig. 4a). For the analysis' sake, the structure is divided in shell strips forming segments of conical frustum in general (Fig 4b). For flanges, these elements degenerate into forms shaped like circular ring-plate segments, and for webs, if they are vertically situated, into strips of a cylindrical shell. In agreement which the character of this variational method and in difference from the folded-plate theory, the trueness of the computation improves with the mesh refinement.

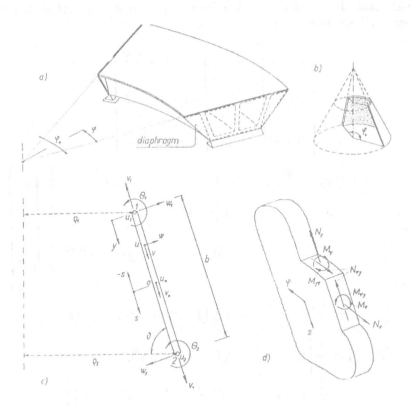

Fig.4.1 *a*. Simply supported curved folded-plate structure having support diaphragms, *b*. structural element in the shape of conical frustum, *c*. cross section of the element and nodal joint degrees of freedom, *d*. internal moments and forces in the shell strip

The finite-strip method differs in two basic points from the folded-plate theory, which, within the limits of the assumptions of the theory of elasticity furnishes an exact solution. In the finite-strip method, supplementary assumptions describing the behaviour of the elements are introduced (in this case these will be the assumptions concerning the distribution of the displacement components along the width of the element), and a different proceeding is adopted to derive the 8-by-8 stiffness matrix for a general conical shell-segment in the n-th mode of harmonic series.

Fig. 4.1c shows the section of an element. The longitudinal edges of this element altogether have eight degrees of freedom which are characterized by the longitudinal displacements u_i, by the displacements v_i in the direction of the surface line, the displacements w_i normal to the surface of the element and by the angles of traverse rotation θ_i of both edges of the element (Fig. 4.1c). The distribution of the in-plane displacements u and v between the longitudinal edges of the element (i. e., along its width) can be considered as a linear variation in the simplest case, the distribution of the displacements w is expressed by the cubic variation. It holds true

$$
\begin{Bmatrix} u(s, \varphi) \\ v(s, \varphi) \\ w(s, \varphi) \end{Bmatrix} = \sum_{n=1}^{\infty} \begin{Bmatrix} u_n(s)\cos\dfrac{n\pi\varphi}{\varphi_0} \\ v_n(s)\sin\dfrac{n\pi\varphi}{\varphi_0} \\ w_n(s)\sin\dfrac{n\pi\varphi}{\varphi_0} \end{Bmatrix} =
$$

$$
= \sum_{n=1}^{\infty} \begin{bmatrix} [F_u(s)]\cos\dfrac{n\pi\varphi}{\varphi_0}, & 0, & 0 \\[2mm] 0, & [F_v(s)]\sin\dfrac{n\pi\varphi}{\varphi_0}, & 0 \\[2mm] 0, & 0, & [F_w(s)]\sin\dfrac{n\pi\varphi}{\varphi_0} \end{bmatrix} \begin{Bmatrix} u_{in} \\ v_{in} \\ w_{in} \end{Bmatrix},
$$

where   (4.1)

$$
[F_u(s)] = \frac{1}{2}[(1-s), (1+s)], \quad [F_v(s)] = \frac{1}{2}[(s-1), (1+s)],
$$

$$
[F_w(s)] = \frac{1}{4}\left[(2 - 3s + s^3), (-2 - 3s + s^3), \right.
$$

$$
\left. \frac{b}{2}(1 - s - s^2 + s^3), \frac{b}{2}(-1 - s + s^2 + s^3) \right] \qquad (4.2)
$$

are the matrices which characterize the polynomials interpolating the respective displacement quantities between their nodal values, b is the width of the element

(Fig. 4.1c), and where

$$\{u_{in}\} = \begin{Bmatrix} u_{1n} \\ u_{2n} \end{Bmatrix}, \qquad \{v_{in}\} = \begin{Bmatrix} v_{1n} \\ v_{2n} \end{Bmatrix}, \qquad \{w_{in}\} = \begin{Bmatrix} w_{1n} \\ w_{2n} \\ \Theta_{1n} \\ \Theta_{2n} \end{Bmatrix}; \qquad (4.3)$$

s is the traverse natural co-ordinate defined to assume (Fig. 4.1c) the value of -1 at the edge 1 and the value of +1 at the edge 2; the natural co-ordinate depends on the co-ordinate y according to the relation $y = 0.5b(s + 1)$. Hence, it holds true for the traverse rotation θ

$$\Theta = \frac{\partial w}{\partial y} = \frac{\partial w}{\partial s} \frac{ds}{dy} = \frac{2}{b} \frac{\partial w}{\partial s}. \qquad (4.4)$$

In order to determine the relations between the generalized strains (strains ε_y, ε_φ and curvatures k_y, k_φ, $k_{y\varphi}$; the index y denotes the direction of the surface line, φ the longitudinal direction) and the displacements of the shell-strip points the following relations are used

$$\varepsilon_y = \frac{\partial v}{\partial y},$$

$$\varepsilon_\varphi = \frac{1}{\varrho} \frac{\partial u}{\partial \varphi} + \frac{\cos \vartheta}{\varrho} v + \frac{\sin \vartheta}{\varrho} w,$$

$$\gamma_{y\varphi} = \frac{\partial u}{\partial y} - \frac{\cos \vartheta}{\varrho} u + \frac{1}{\varrho} \frac{\partial v}{\partial \varphi},$$

$$k_y = -\frac{\partial^2 w}{\partial y^2}, \qquad\qquad\qquad (4.5)$$

$$k_\varphi = \frac{\sin \vartheta}{\varrho^2} \frac{\partial u}{\partial \varphi} - \frac{1}{\varrho^2} \frac{\partial^2 w}{\partial \varphi^2} - \frac{\cos \vartheta}{\varrho} \frac{\partial w}{\partial y},$$

$$k_{y\varphi} = \frac{2 \sin \vartheta}{\varrho} \frac{\partial u}{\partial y} - \frac{2 \sin \vartheta \cos \vartheta}{\varrho^2} u - \frac{2}{\varrho} \frac{\partial^2 w}{\partial y \partial \varphi} + \frac{2 \cos \vartheta}{\varrho^2} \frac{\partial w}{\partial \varphi},$$

where v is the gradient of the shell strip (Fig. 4.1c) and ρ is the radial co-ordinate,

$$\varrho = \frac{\varrho_2 + \varrho_1}{2} + \frac{\varrho_2 - \varrho_1}{2} s.$$

Then the relations (4.1) are substituted into the relations (4.5), with regard to the expressions (4.2) and (4.3), and the expression of the strain vector in dependence on the displacements is obtained. It holds true for the n-th harmonic

$$
\begin{Bmatrix} \varepsilon_{yn} \\ \varepsilon_{\varphi n} \\ \gamma_{y\varphi n} \\ k_{yn} \\ k_{\varphi n} \\ k_{y\varphi n} \end{Bmatrix} =
\begin{bmatrix}
0, & \dfrac{dF_v}{dy}\sin\dfrac{n\pi\varphi}{\varphi_0}, \\[2ex]
-\dfrac{n\pi}{\varrho\varphi_0}F_u\sin\dfrac{n\pi\varphi}{\varphi_0}, & \dfrac{\cos\vartheta}{\varrho}F_v\sin\dfrac{n\pi\varphi}{\varphi_0}, \\[2ex]
\left(\dfrac{dF_u}{dy}-\dfrac{\cos\vartheta}{\varrho}F_u\right)\cos\dfrac{n\pi\varphi}{\varphi_0}, & \dfrac{n\pi}{\varrho\varphi_0}F_v\cos\dfrac{n\pi\varphi}{\varphi_0}, \\[2ex]
0, & 0, \\[2ex]
-\dfrac{\sin\vartheta}{\varrho^2}\dfrac{n\pi}{\varphi_0}F_u\sin\dfrac{n\pi\varphi}{\varphi_0}, & 0, \\[2ex]
2\left(\dfrac{\sin\vartheta}{\varrho}\dfrac{dF_u}{dy}-\dfrac{\sin\vartheta\cos\vartheta}{\varrho^2}F_u\right)\cos\dfrac{n\pi\varphi}{\varphi_0}, & 0,
\end{bmatrix}
$$

$$
\begin{bmatrix}
0 \\[2ex]
\dfrac{\sin\vartheta}{\varrho}F_w\sin\dfrac{n\pi\varphi}{\varphi_0} \\[2ex]
0 \\[2ex]
-\dfrac{d^2F_w}{dy^2}\sin\dfrac{n\pi\varphi}{\varphi_0} \\[2ex]
\left[\left(\dfrac{n\pi}{\varrho\varphi_0}\right)^2 F_w-\dfrac{\cos\vartheta}{\varrho}\dfrac{dF_w}{dy}\right]\sin\dfrac{n\pi\varphi}{\varphi_0} \\[2ex]
\dfrac{2n\pi}{\varrho\varphi_0}\left(\dfrac{\cos\vartheta}{\varrho}F_w-\dfrac{dF_w}{dy}\right)\cos\dfrac{n\pi\varphi}{\varphi_0}
\end{bmatrix}
\begin{Bmatrix} u_{in} \\ v_{in} \\ w_{in} \end{Bmatrix} \quad (4.6)\;)
$$

This can be written in the form of

$$\{\varepsilon_n\} = [Z_n]\{\delta_n\}. \tag{4.7}$$

The relation between the generalized internal forces, and/or the moments in the shell strip and the strain vector, when orthogonal anisotropic material is considered, is

$$
\begin{Bmatrix} N_y \\ N_\varphi \\ N_{y\varphi} \\ M_y \\ M_\varphi \\ M_{y\varphi} \end{Bmatrix}
=
\begin{bmatrix}
\dfrac{{}^s t\, {}^s E_y}{1 - {}^s v_{y\varphi}\,{}^s v_{\varphi y}}, & {}^s v_{\varphi y}\dfrac{{}^s t\, {}^s E_\varphi}{1 - {}^s v_{y\varphi}\,{}^s v_{\varphi y}}, & 0, & 0, & 0, & 0 \\[3mm]
{}^s v_{y\varphi}\dfrac{{}^s t\, {}^s E_y}{1 - {}^s v_{y\varphi}\,{}^s v_{\varphi y}}, & \dfrac{{}^s t\, {}^s E_\varphi}{1 - {}^s v_{y\varphi}\,{}^s v_{\varphi y}}, & 0, & 0, & 0, & 0 \\[3mm]
0, & 0, & {}^s t\, {}^s G, & 0, & 0, & 0 \\[3mm]
0, & 0, & 0, & \dfrac{{}^d t^3\, {}^d E_y}{12(1 - {}^d v_{y\varphi}\,{}^d v_{\varphi y})}, & {}^d v_{\varphi y}\dfrac{{}^d t^3\, {}^d E_\varphi}{12(1 - {}^d v_{y\varphi}\,{}^d v_{\varphi y})}, & 0 \\[3mm]
0, & 0, & 0, & {}^d v_{y\varphi}\dfrac{{}^d t^3\, {}^d E_y}{12(1 - {}^d v_{y\varphi}\,{}^d v_{\varphi y})}, & \dfrac{{}^d t^3\, {}^d E_\varphi}{12(1 - {}^d v_{y\varphi}\,{}^d v_{\varphi y})}, & 0 \\[3mm]
0, & 0, & 0, & 0, & 0, & \dfrac{{}^d t^3\, {}^d G}{12}
\end{bmatrix}
\begin{Bmatrix} \varepsilon_y \\ \varepsilon_\varphi \\ \gamma_{y\varphi} \\ k_y \\ k_\varphi \\ k_{y\varphi} \end{Bmatrix},
\tag{4.8}
$$

that is

$$\{\sigma_n\} = [D]\{\varepsilon_n\}, \tag{4.9}$$

where N_y, N_φ are the specific normal internal forces, N_y the specific shear force, M_φ, M_y the specific bending moments, and $M_{y\varphi}$ the specific torsion moment in the shell strip (Fig. 4.1d). The orthotropy of the material is characterized by the moduli of elasticity and by the Poisson's ratio, differentiated by the indices y and φ according to the direction in which these material constants are in force. The left-hand upper indices s and d denote the pertinence to the membrane (s) and bending (d) actions of the element. The matrix $[D]$ may be also of constitutive nature, all its independent elements can be specified directly, thus completely defining a plate type. Stiffeners in either direction and different amounts of reinforcement can be taken into account in this way. The d_{jk} coefficients can be determined, if necessary, also experimentally.

By combining the relations (4.7) and (4.9) it can be written

$$\{\sigma_n\} = [D] [Z_n] \{\delta_n\}, \tag{4.10}$$

which explicitly gives Eq. (4.11) where d_{jk} are the elements of the matrix $[D]$ [see relation (4.8)].

The potential energy stored in a shell strip can be expressed by using Eqs. (4.7) and (4.10) as

$$A = \frac{1}{2} \int_\Omega \{\varepsilon\}^T \{\sigma\}\, d\Omega =$$

$$= \frac{1}{2} \sum_{n=1}^{\infty} \sum_{m=1}^{\infty} \{\delta_n\}^T (\int_{y=0}^{b} \int_{\varphi=0}^{\varphi_0} [Z_n]^T [D] [Z_m]\, \varrho\, dy\, d\varphi)\{\delta_m\}, \tag{4.12}$$

i. e., in dependence on the joint displacements and the geometrical and physical properties of the shell strip.

The products indicated in Eq. (4.12) for $n \neq m$ vanish for reasons of orthogonality of the displacement functions, and for $n = m$ it holds

$$\int_0^{\varphi_0} \sin^2 \frac{n\pi\varphi}{\varphi_0}\, d\varphi = \int_0^{\varphi_0} \cos^2 \frac{n\pi\varphi}{\varphi_0}\, d\varphi = \frac{\varphi_0}{2}. \tag{4.13}$$

Eq. (4.12) is thus simplified to

$$A = \frac{1}{2} \sum_{n=1}^{\infty} \{\delta_n\}^T \left(\frac{\varphi_0}{2} \int_{y=0}^{b} [\overline{Z}_n]^T [D] [\overline{Z}_n]\, \varrho\, dy \right) \{\delta_n\}, \tag{4.14}$$

where the matrix $[\overline{Z}_n]$ is equal to the matrix $[Z_n]$ with all the multipliers deleted, in other words, it is a matrix whose elements are the amplitudes of the elements of the matrix $[Z_n]$.

From Eq. (4.14) follows the relation for the stiffness matrix for a typical harmonic n

$$[k_n] = \frac{\varphi_0}{2} \int_0^b [\overline{Z}_n]^T [D] [\overline{Z}_n]\, \varrho\, dy = \frac{\varphi_0 b}{4} \int_{-1}^1 [\overline{Z}_n]^T [D] [\overline{Z}_n]\, \varrho\, ds. \tag{4.15}$$

This stiffness matrix of the size 8 by 8 can be, with regard to the displacement components u, v and w, partitioned in fields

$$[k_n] = \frac{\varphi_0 b}{4} \begin{bmatrix} [k_{uun}] & [k_{uvn}] & [k_{uwn}] \\ [k_{uvn}] & [k_{vvn}] & [k_{vwn}] \\ [k_{uwn}] & [k_{vwn}] & [k_{wwn}] \end{bmatrix}, \tag{4.16}$$

$$
\begin{Bmatrix} N_{yn} \\ N_{\varphi n} \\ N_{y\varphi n} \\ M_{yn} \\ M_{\varphi n} \\ M_{y\varphi n} \end{Bmatrix}
=
\left[
\begin{array}{cc}
-\dfrac{n\pi}{\varrho\varphi_0}d_{12}F_u\sin\dfrac{n\pi\varphi}{\varphi_0}, & \left(d_{11}\dfrac{\mathrm{d}F_v}{\mathrm{d}y}-d_{12}\dfrac{\cos\vartheta}{\varrho}F_v\right)\sin\dfrac{n\pi\varphi}{\varphi_0}, \\[2mm]
-\dfrac{n\pi}{\varrho\varphi_0}d_{22}F_u\sin\dfrac{n\pi\varphi}{\varphi_0}, & \left(d_{12}\dfrac{\mathrm{d}F_v}{\mathrm{d}y}+d_{22}\dfrac{\cos\vartheta}{\varrho}F_v\right)\sin\dfrac{n\pi\varphi}{\varphi_0}, \\[2mm]
d_{33}\left(\dfrac{\mathrm{d}F_u}{\mathrm{d}y}-\dfrac{\cos\vartheta}{\varrho}F_u\right)\cos\dfrac{n\pi\varphi}{\varphi_0}, & \dfrac{n\pi}{\varrho\varphi_0}d_{33}F_v\cos\dfrac{n\pi\varphi}{\varphi_0}, \\[2mm]
-\dfrac{n\pi}{\varrho\varphi_0}\dfrac{\sin\vartheta}{\varrho}d_{45}F_u\sin\dfrac{n\pi\varphi}{\varphi_0}, & 0, \\[2mm]
-\dfrac{n\pi}{\varrho\varphi_0}\dfrac{\sin\vartheta}{\varrho}d_{55}F_u\sin\dfrac{n\pi\varphi}{\varphi_0}, & 0, \\[2mm]
\dfrac{\sin^2\vartheta}{\varrho}d_{66}\left(\dfrac{\mathrm{d}F_u}{\mathrm{d}y}-\dfrac{\cos\vartheta}{\varrho}F_u\right)\cos\dfrac{n\pi\varphi}{\varphi_0}, & 0,
\end{array}
\right.
$$

$$
\left.
\begin{array}{c}
\dfrac{\sin\vartheta}{\varrho}d_{12}F_w\sin\dfrac{n\pi\varphi}{\varphi_0} \\[2mm]
\dfrac{\sin\vartheta}{\varrho}d_{22}F_w\sin\dfrac{n\pi\varphi}{\varphi_0} \\[2mm]
0 \\[2mm]
\left[\left(\dfrac{n\pi}{\varrho\varphi_0}\right)^2 d_{45}F_w - d_{44}\dfrac{\mathrm{d}^2F_w}{\mathrm{d}y^2}-\dfrac{\cos\vartheta}{\varrho}d_{45}\dfrac{\mathrm{d}F_w}{\mathrm{d}y}\right]\sin\dfrac{n\pi\varphi}{\varphi_0} \\[2mm]
\left[\left(\dfrac{n\pi}{\varrho\varphi_0}\right)^2 d_{55}F_w - d_{45}\dfrac{\mathrm{d}^2F_w}{\mathrm{d}y^2}-\dfrac{\cos\vartheta}{\varrho}d_{55}\dfrac{\mathrm{d}F_w}{\mathrm{d}y}\right]\sin\dfrac{n\pi\varphi}{\varphi_0} \\[2mm]
2d_{66}\dfrac{n\pi}{\varrho\varphi_0}\left(\dfrac{\cos\vartheta}{\varrho}F_w-\dfrac{\mathrm{d}F_w}{\mathrm{d}y}\right)\cos\dfrac{n\pi\varphi}{\varphi_0}
\end{array}
\right]
\begin{Bmatrix} u_{in} \\ v_{in} \\ w_{in} \end{Bmatrix},
$$

$$(4.11)$$

where the individual fields are given by the matrices

$$[k_{uun}] = \left[d_{22}\left(\frac{n\pi}{\varphi_0}\right)^2 + d_{33}\cos^2\vartheta\right]\int_{-1}^{1}\frac{1}{\varrho}[F_u]^T[F_u]\,ds +$$

$$+ \sin^2\vartheta\left[d_{55}\left(\frac{n\pi}{\varphi_0}\right)^2 + 4d_{66}\cos^2\vartheta\right]\int_{-1}^{1}\frac{1}{\varrho^3}[F_u]^T[F_u]\,ds -$$

$$- d_{33}\cos\vartheta\int_{-1}^{1}\left(\left[\frac{dF_u}{dy}\right]^T[F_u] + [F_u]^T\left(\frac{dF_u}{dy}\right)\right)ds -$$

$$- 4d_{66}\sin^2\vartheta\cos\vartheta\int_{-1}^{1}\frac{1}{\varrho^2}\left(\left[\frac{dF_u}{dy}\right]^T[F_u] + [F_u]^T\left[\frac{dF_u}{dy}\right]\right)ds +$$

$$+ d_{33}\int_{-1}^{1}\left[\frac{dF_u}{dy}\right]^T\left[\frac{dF_u}{dy}\right]\varrho\,ds + 4d_{66}\sin^2\vartheta\int_{-1}^{1}\frac{1}{\varrho}\left[\frac{dF_u}{dy}\right]^T\left[\frac{dF_u}{dy}\right]ds,$$

$$[k_{uvn}] = -\frac{n\pi}{\varphi_0}\cos\vartheta(d_{22} + d_{33})\int_{-1}^{1}\frac{1}{\varrho}[F_u]^T[F_v]\,ds +$$

$$+ \frac{n\pi}{\varphi_0}d_{33}\int_{-1}^{1}\left[\frac{dF_u}{dy}\right]^T[F_v]\,ds - \frac{n\pi}{\varphi_0}d_{12}\int_{-1}^{1}[F_u]^T\left[\frac{dF_u}{dy}\right]ds,$$

$$[k_{uwn}] = -\frac{n\pi}{\varphi_0}d_{22}\sin\vartheta\int_{-1}^{1}\frac{1}{\varrho}[F_u]^T[F_w]\,ds -$$

$$- \sin\vartheta\frac{n\pi}{\varphi_0}\left[d_{55}\left(\frac{n\pi}{\varphi_0}\right)^2 + 4d_{66}\cos^2\vartheta\right]\int_{-1}^{1}\frac{1}{\varrho^3}[F_u]^T[F_w]\,ds +$$

$$+ \sin\vartheta\cos\vartheta\frac{n\pi}{\varphi_0}(d_{55} + 4d_{66})\int_{-1}^{1}\frac{1}{\varrho^2}[F_u]^T\left[\frac{dF_w}{dy}\right]ds +$$

$$+ \sin\vartheta\frac{n\pi}{\varphi_0}d_{45}\int_{-1}^{1}\frac{1}{\varrho}[F_u]^T\left[\frac{d^2F_w}{dy^2}\right]ds + 4d_{66}\sin\vartheta\times \qquad (4.17)$$

$$\times\left(\cos\vartheta\frac{n\pi}{\varphi_0}\int_{-1}^{1}\frac{1}{\varrho^2}\left[\frac{dF_u}{dy}\right]^T[F_w]\,ds - \frac{n\pi}{\varphi_0}\int_{-1}^{1}\frac{1}{\varrho}\left[\frac{dF_u}{dy}\right]^T\left[\frac{dF_w}{dy}\right]ds\right),$$

$$[k_{vvn}] = \left[d_{22}\cos^2\vartheta + d_{33}\left(\frac{n\pi}{\varphi_0}\right)^2\right]\int_{-1}^{1}\frac{1}{\varrho}[F_v]^T[F_v]\,ds +$$

$$+ d_{12}\cos\vartheta\int_{-1}^{1}\left(\left[\frac{dF_v}{dy}\right]^T[F_v] + [F_v]^T\left[\frac{dF_v}{dy}\right]\right)ds +$$

$$+ d_{11}\int_{-1}^{1}\left[\frac{dF_v}{dy}\right]^T\left[\frac{dF_v}{dy}\right]\varrho\,ds,$$

$$[k_{vwn}] = d_{22} \sin \vartheta \cos \vartheta \int_{-1}^{1} \frac{1}{\varrho} [F_v]^T [F_w] ds + d_{12} \sin \vartheta \int_{-1}^{1} \left[\frac{dF_v}{dy} \right]^T [F_w] ds,$$

$$[k_{wwn}] = d_{22} \sin^2 \vartheta \int_{-1}^{1} \frac{1}{\varrho} [F_w]^T [F_w] ds +$$

$$+ \left(\frac{n\pi}{\varphi_0} \right)^2 \left[d_{55} \left(\frac{n\pi}{\varphi_0} \right)^2 + 4 d_{66} \cos^2 \vartheta \right] \int_{-1}^{1} \frac{1}{\varrho^3} [F_w]^T [F_w] ds -$$

$$- (d_{55} + 4 d_{66}) \cos \vartheta \left(\frac{n\pi}{\varphi_0} \right)^2 \int_{-1}^{1} \frac{1}{\varrho^2} \left(\left[\frac{dF_w}{dy} \right]^T [F_w] + [F_w]^T \left[\frac{dF_w}{dy} \right] \right) ds -$$

$$- d_{45} \left(\frac{n\pi}{\varphi_0} \right)^2 \int_{-1}^{1} \frac{1}{\varrho} \left([F_w]^T \left[\frac{d^2 F_w}{dy^2} \right] + \left[\frac{d^2 F_w}{dy^2} \right]^T [F_w] \right) ds +$$

$$+ \left[d_{55} \cos^2 \vartheta + 4 d_{66} \left(\frac{n\pi}{\varphi_0} \right)^2 \right] \int_{-1}^{1} \frac{1}{\varrho} \left[\frac{dF_w}{dy} \right]^T \left[\frac{dF_w}{dy} \right] ds +$$

$$+ d_{45} \cos \vartheta \int_{-1}^{1} \left(\left[\frac{dF_w}{dy} \right]^T \left[\frac{d^2 F_w}{dy^2} \right] + \left[\frac{d^2 F_w}{dy^2} \right]^T \left[\frac{dF_w}{dy} \right] \right) ds +$$

$$+ d_{44} \int_{-1}^{1} \left[\frac{d^2 F_w}{dy^2} \right]^T \left[\frac{d^2 F_w}{dy^2} \right] \varrho \, ds.$$

Numerical calculations conducted according to these relations are very sensitive to numerical accuracy; for large radii of curvature in particular (i. e., for nearly straight structures), it is advantageous to use some less sensitive numerical procedure instead of a direct computation of the indicated integrals.

In this way the stiffness matrices of all the elements of the structure (shell strips) are formed. Further proceeding of the analysis is fully coincident with the analysis of folded-plate structures: the stiffness matrices of the elements are transformed into a global co-ordinate system by using transformation matrices which are of the same nature as those given, by Eq. (2.35).

Finally, by assembling the transformed stiffness matrices, the stiffness matrix of the structure is formed; the effect of an external loading of the structure is described by the right-hand sides; systems of equations of the type (2.45) for each n-th harmonic are established. The ascertained final joint displacements, after a reverse transformation into the local co-ordinate system, are introduced into the relations (4.10) or (4.11) respectively, and the final internal forces in the structure are accumulated as the sum of the harmonic contributions.

References

1 Timoshenko, S.P. and Gere, J.M.: *Theory of Elastic Stability*, McGraw-Hill Book Co, New York, 1961
2 Vlasov, V.Z.: *Thin-Walled Elastic Beams*, US Dept. of Commerce, PST Catalogue 428, 1959
3 Bleich, F.: *Buckling Strength of Metal Structures*, McGraw-Hill, New York, 1952
4 Umanskij, A.A.: *Torsion and Bending of Thin-Walled Aerostructures*, (in Russian), Oborogiz, Moscow, 1939
5 Kollbrunner, C.F. and Hajdin: *Dunnwandige Stabe, Springer-Verlag*, Berlin,Heidelberg, New York, 1975
6 Bažant, Z.P. and El Nimeiri: *Stiffness Method for Curved Box Girders at Initial Stress*, Journal of the Structural Division, ASCE, , 100(10), 2071-90, 1974
7 Křístek, V.: *Tapered Box Girders of Deformable Cross Section*, Journal of the Structural Division, ASCE Proc. Paper 7489, August 1970
8 Křístek, V.: *Theory of Box Girders*, John Wiley and Sons, Chichester, New York, Brisbane, Toronto, 1979
9 Maisel, B.I.: *Methods of Analysis and Design of Concrete Boxbeams with Side Cantilevers*, Technical Report, Cement and Concrete Association, London,1974
10 Megson, T.H.G.: *Linear Analysis of Thin-Walled Elastic Structures*, John Wiley and Sons, New York, 1974
11 Goldberg, J.E. and Leve, H.L.: *Theory of Prismatic Folded Plate Structures*, Publ.International Association for Bridge and Structural Engineering, Vol. 17, 1957
12 Křístek, V.: *Folded Plates with Deformable End Cross Sections*, Proc. Instn. Civ Engrs, Part 2, London, September 1983
13 Meyer, C.: *Analysis and Design of Curved Box Girders*, University of California, Berkeley, 1970

SHEAR LAG IN WIDE FLANGES
AND
THE "BREATHING" OF SLENDER WEB PLATES

M. Skaloud
Czech Academy of Sciences, Prague, Czech Republic

Abstract

This part of the monograph, corresponding to six hours of lecturing at the related Advanced School, deals with two chapters on advanced analysis of steel plate and box girders; viz. (i) shear lag in wide flanges and (ii) the „breathing" of slender web plates. Both chapters are mostly based on the results and conclusions of the research undertaken by the author and his associates in Prague during the last years.

1. Shear Lag in Wide Flanges

In a box girder (Fig.1.1 a), the web and flange plates are interconnected so that relative displacements cannot occur. Therefore, at the junction of the web with the flange the longitudinal strain in the web $(\varepsilon_{x,w})$ must be equal to that in the flange $(\varepsilon_{x,f})$. A shear flow develops between the web and the flange which causes

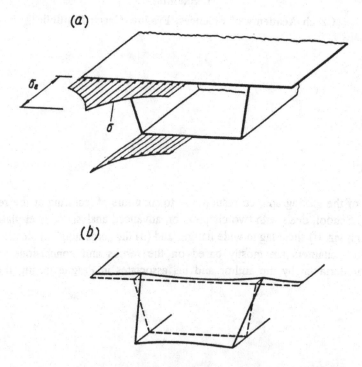

Fig. 1.1 Shear lag effects: (a) distribution of longitudinal normal stresses across flange widths, (b) warping of the cross-section.

shear deformation of the flange plate. The longitudinal displacements in the parts of the flange remote from the webs lag behind those nearer the webs (Fig. 1.1 b). This effect leads to a non-uniform distribution of the longitudinal normal stresses across the flange width (Fig. 1.1a). The effect is particularly pronounced in wide flanges and in flanges with longitudinal stiffeners.

This phenomenon, termed shear lag, results in a considerable increase of the longitudinal stresses σ in the regions of the flange close to the webs in comparison with those given by the elementary theory of bending (Fig. 1.1 a). Thus neglect of shear lag would lead to an underestimation of the stresses developed in the flange plates at positions adjacent to the webs, and hence to an unsafe design. The shear lag may also significantly influence the girder deflections.

It is an accepted practice in structural engineering to represent the effect of shear lag by adopting an effective breadth concept. The actual width of the flange plate b is replaced by a reduced width b_{ef} over which the longitudinal stresses may be considered uniformly distributed, and the application of the elementary theory of bending to the transformed girder cross-section gives the correct value of a maximum longitudinal stress σ_e (Fig. 1.1 a). A similar procedure may also be carried out for deflections. However, when the structure is subjected to large concentrated loads, the concept of effective breadth gives reliable information only in those parts of the structure that are not very close to the point of application of the load or the support reaction. In the immediate neighbourhood of a point load, the actual stress state can differ rather substantially from that resulting from any simplified analysis, including the effective breadth concept.

1.1   Methods of Analysis

Extensive analysis of the shear lag effect has been carried out during the last few decades. An analytical method has been given by Girkmann [1.2]. More recently many analytical models and methods have been developed. Among these are numerical solutions based on finite element or finite difference methods, exact and approximate methods based on folded plate theory, and approximate methods based on simplified structural behaviour.

1.1.1   The Finite Element Method

The finite element method has become practically universal for the solution of mechanics problems in recent years. The continuum is replaced by an assembly of finite elements interconnected at nodal points. Stiffness matrices are developed for the finite elements based on assumed displacement patterns, and then an analysis based on the direct stiffness method may be performed to determine nodal point displacements and, subsequently, the internal stresses in the finite elements. As this method is well documented in the literature, no attempt is made to review it here in detail.

Moffat and Dowling [1.3] produced a comprehensive parametric study of the shear lag effect in box girders; this study was based on the use of the finite element method. They found that while only one mesh division over the depth of a girder was sufficient, fine mesh divisions had to be used over the girder width and length, particularly in the region of a point load or a support.

Although a finite element solution is capable of giving a comprehensive and adequate picture of the stress distribution, it requires the use of large computers and is too costly, particularly if repeated analyses are required at the preliminary design stage.

1.1.2   The Folded Plate Theory

Steel box girders are usually of constant cross-section and, hence, the folded plate theory is ideally suited to predicting shear lag effects. This theory; described fully above in [1.1], takes advantage of harmonic analysis and may be applied to a variety of support conditions. Use of the folded plate theory results in a considerable saving of computer time over the finite element method.

1.1.3   The Finite Strip Method

A direct application of the theory of elasticity to determine the stiffness matrix of curved folded plate elements becomes exceedingly complex. A theory known as the finite strip method [1.4], [1.5], [1.6] may be used in these cases. This method may be considered as a special form of the finite element method. It approximates the behaviour of each plate by an assembly of longitudinal finite strips for which selected displacement patterns, varying as harmonics longitudinally and as polynomials in the transverse direction, are assumed to represent the behaviour of the strip in the total structure. With this assumption, the displacement at any point in the strip can be expressed in terms of eight nodal point displacements and, hence, the element stiffness matrix determined. The remaining procedure is similar to that used in the folded plate method.

1.1.4   Harmonic Analysis of Shear Lag in Flanges with Closely-Spaced Stiffeners and in Composite Flanges

A simple method, employing harmonic analysis, enables shear lag effects in wide flanges to be predicted from hand calculations [1.1], [1.7].

The method is suitable for the analysis of girders with flanges which are not stiffened, as in the case of concrete girders or flanges of composite girders, and of girders where the stiffeners are so closely spaced that it is reasonable to assume the stiffener properties to be spread evenly (or smeared) over the flange width. This is indeed the case for many bridge and aircraft girders. The method may be applied to multi-cellular girders, as well as to girders with inclined web plates and with overhanging or cantilevered flanges.

Although the method is suitable for hand calculations, it has been programmed for a personal computer for added convenience; a suitable program is included in [1.1].

1.2   The Main Features of Shear Lag and the Influence of Various Parameters

1.2.1   Variation of Width/Span Ratio

It is well known that as the flange width increases in relation to the span, the shear lag effect becomes more pronounced. This finding is of general validity and can be clearly illustrated.

The results of such a parametric study are shown in Fig.1.2, where the edge stress σ_e for a box girder under distributed loading is plotted against the girder span. The calculated stress is compared to the stress σ_0 predicted by simple beam theory.

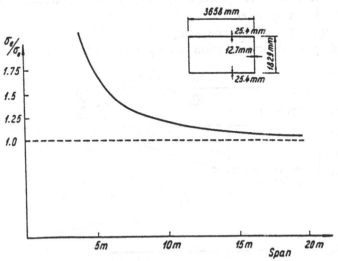

Fig. 1.2. Influence of span length upon the shear lag effect.

1.2.2   The Type and Position of Loading

The non-uniformity of distribution of longitudinal stresses increases rapidly in the region of a point load or a support (see also Section 1.2.4). Moving the load system away from the mid-span results in a reduction of the effective breadth ratios (Moffat and Dowling [1.3]). The effective breadth ratios are only sensitive to the loaded length if this length is less than half of the span.

1.2.3   Effect of Stiffeners

Because stiffeners contribute to the axial load-carrying capacity of a flange without increasing its shear capacity, shear lag is more pronounced in a stiffened flange than in a flange without stiffening. As an example, Fig. 1.3 shows the results of a parametric study investigating the influence of the variations of the ratio $\bar{t}/t$ in a steel girder (without any concrete layer) where $\bar{t} = t + A_s/a$. The unstiffened steel girder is represented by the case when $\bar{t} = t$ and the ratio is then increased to a value of 2, representing a heavily stiffened girder. Within this range, the stress at the edge of the flange is seen to increase almost linearly.

For practical purposes, it is desirable to simplify the numerical analysis by smearing the stiffener properties over the flange width. This approach must be verified from the point of view of:

(i) the number of longitudinal ribs;

(ii) the effect of their own flexural rigidities and the eccentricity of their connections to the flange sheet;

(iii) the regularity of stiffener spacing;

(iv) the shape of the cross-section of longitudinal ribs.

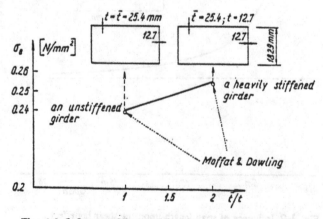

Fig. 1.3. Influence of flange stiffening upon the shear lag effect.

It is obvious that with an increase in the number of stiffeners, the moment of inertia of the whole cross-section also increases and, consequently, the magnitude of the stress drops. It has been found, however that, in spite of the influence of the rate of flange stiffening discussed above, the general character of the stress distribution remains similar. This is clearly seen from Fig.1.3, which, as an example, gives the distribution of longitudinal normal stresses over the flange breadth for various numbers of longitudinal flat ribs (in all cases the cross-sectional area of the stiffeners being the same). Hence, it seems that the concept of smearing stiffener properties is acceptable not only for closely-spaced stiffeners but also, with little loss of accuracy, for large, rather widely-spaced stiffeners. The conditions for the acceptability of such an approach are the regularity of the stiffener arrangement and the assumption that the stiffeners are concentrated in the flange plane (Fig.1.5a).

Longitudinal stiffeners are generally welded to the inner side of the flange sheet (Fig. 1.5 b). Due to the eccentricity of the stiffener connection, individual portions of the flange with stiffeners, which are eccentrically affected by shear flows acting in the plane of the flange sheet (Fig. 1.5 c), tend to exhibit additional flexure, as shown in Fig. 1.6 d.

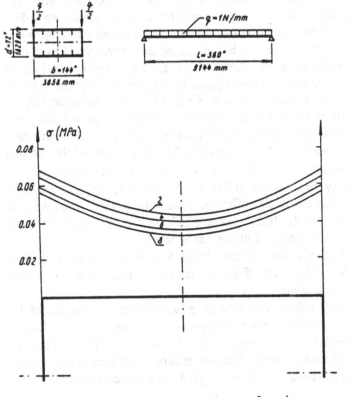

Fig. 1.4. Distribution of longitudinal normal stresses for various numbers of longitudinal flat ribs.

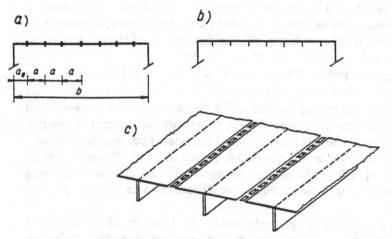

Fig. 1.5. Stiffeners concentrated at the flange sheet plane
and eccentrically-connected stiffeners.

The influence of eccentricity of the stiffener connection is illustrated in Fig.
1.6. A steel box girder without intermediate diaphragms, under uniform loading $(w = 1 \text{ N/mm})$, with span $L = 9144$ mm and the cross-section shown in Fig.
1.6 a is studied. The distribution of longitudinal stresses across the flange width
is shown by the solid line in Fig. 1.6 b; the dashed line corresponds to that of
the stiffeners concentrated in the flange sheet. Different flexural actions of
individual stiffeners with adjacent flange portions are shown in Fig. 1.6 c, which
depicts distributions of the longitudinal stresses along the stiffener depths.

It can be seen from Fig. 1.6 c that the eccentrically connected stiffeners,
particularly those near the mid-point of the flange, exhibit stress distribution
tending to that of a beam stressed by bending. The eccentricity of the stiffener
connections thus results in a loss of efficiency of the total stiffened flange. This
is also the reason why the solid line in Fig. 1.6 b, indicating the stress distribution for the eccentrically connected stiffeners, falls completely (i.e. along the total
width of the flange) above the dashed curve, which corresponds to a fully acting
flange with stiffeners concentrated at the flange sheet.

The transverse flexure of the stiffened flange due to stiffener eccentricity is
shown in Fig. 1.6 d. This kind of additional deformation is only partially
restrained by a rather flexible flange.

It is seen that the stiffener eccentricity influences the distribution of the
longitudinal stresses adversely, unless closely-spaced sufficiently rigid transverse
diaphragms are used to ensure equal deflection of all stiffeners. Thus the
diaphragms indirectly influence the shear lag effects. Their presence is essential
to allow use of the methods that do not regard stiffener eccentricity.

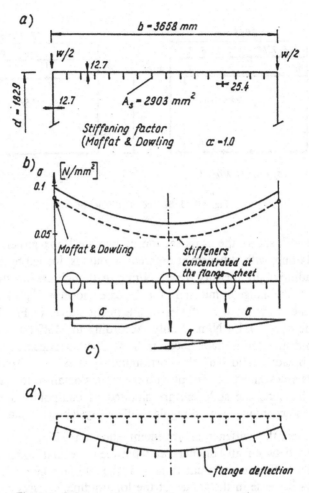

Fig. 1.6. Influence of eccentrically-connected stiffeners upon
the stress distribution and deflections.

It has been found that the regularity of stiffener spacing (even in cases of the
same total cross-sectional area of stiffeners and thus the same total second
moment of area of the whole cross-section) influence the shear lag behaviour of
the stiffened plate.

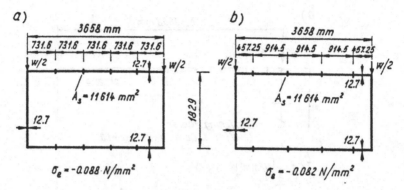

Fig. 1.7. Influence of stiffener positions.

Figure 1.7 shows the cross-section of two steel box girders having a span $L = 9\,144$ mm, with stiffeners at different positions, loaded by uniformly distributed loading of an intensity $w = 1$ N/mm. Although the longitudinal stress on the edge of the flange at mid-span for the case shown in Fig. 1.7'a is $-0.087\,55$ N/mm^2, the stress for the stiffener arrangement shown in Fig. 3.17b reaches a magnitude of $-0.081\,8$ N/mm^2 only. According to Moffat and Dowling [1.3], where no distinction is made between the stiffener arrangements, for the (stress) effective breadth ratio 0.67 the corresponding stress is -0.082 N/mm^2. This represents an excellent agreement with the results obtained for the case shown in Fig. 1.7'b. Here, the stiffeners are situated at mid-points of adjacent flange portions in accordance with a regular stiffener system as assumed in [1.3].

However, the stiffener arrangement shown in Fig. 1.7a (with the same distances between all stiffeners and between the first stiffener and the web), where the stiffeners are situated more at the middle region of the flange, results in a 7 % increase in the values of the longitudinal stresses compared with the case shown in Fig. 1.7'b. The reason for this is that the shear lag effect depends on shear deformability of those flange segments where the shear stress is of highest intensity, i.e. in the regions close to the webs. The width of the flange segment between the web and the first stiffener (and thus its shear deformability) is considerably lower in the case shown in Fig. 1.7'b than in that shown in Fig. 1.7'a.

The results clearly confirm the necessity of accounting for the actual stiffener positions for flanges with large, widely spaced stiffeners, e.g. by using the method presented in Section [1.1]. This fact cannot be accounted for by any method of analysis which assumes a regularly arranged structure, even if the finite element method is used.

Most studies of the impact of the shear lag phenomenon characterize longitudinal ribs solely by their area, and the effect of stiffener configuration is not taken into account.

Investigations dealing with the stability problem of longitudinally stiffened flanges, in [1.1], proved the great influence of stiffener cross-sectional shape upon the buckling of flange plates. Thus it is of interest to find out whether the stiffener cross-section configuration also shows a significant effect on the shear lag phenomenon.

For this reason, the effect of various stiffener shapes was investigated, while other stiffener parameters (the number and location of stiffeners, and – at least approximately – the moment of inertia) were kept constant. Four stiffener configurations were studied: (i) a flat stiffener (Fig. 1.8 a), (ii) an angle stiffener (Fig. 1.8 b), (iii) a T-section stiffener (Fig. 1.8 c) and (iv) a trapezoidal closed-section stiffener (as shown in Fig. 1.8 d). The overall dimensions of the girders analysed are shown in Fig. 1.4 .

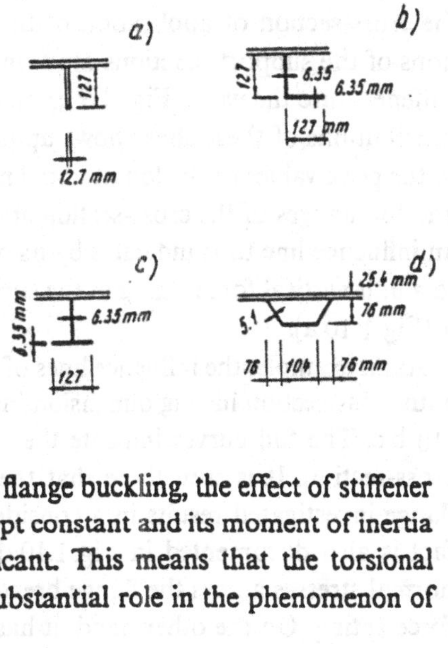

Fig. 1.8. The stiffener configurations considered in the study.

The results show that, unlike the case of flange buckling, the effect of stiffener configuration (when the stiffener area is kept constant and its moment of inertia does not vary much either) is not significant. This means that the torsional rigidity of the stiffeners does not play a substantial role in the phenomenon of shear lag.

1.2.4   Shear Lag in Support Regions of Continuous Beams

When analysing the shear lag in box girders, the differences in performance of the top and bottom flanges are usually tacitly ignored in practice. It has been found, however, that this concept gives satisfactory results only for box girders of ordinary dimensions (i.e. for girders which are not too short) which are subjected to uniformly distributed and similar loadings. In reality, the external loading is usually first transmitted by a system of transverse ribs to the upper edges of webs, while on the other hand the lower edges of the webs are subjected to support reactions representing very large point loads (Fig. 1.9 a,b)

The webs of box girders are usually reinforced by vertical stiffeners located at the cross-section of application of the reaction, Fig. 1.9 a,b. Substantial portions of the support reactions are transmitted (as shear flows) by these vertical stiffeners into the webs (Fig. 1.9 c). In order to evaluate the influence of various distributions of these shear flows upon the shear lag effects, the influence lines of the peak values of the longitudinal normal stresses at the edges of the bottom and top flanges of the cross-section above the support can be constructed. Such an influence line thus indicates by its ordinates the value of the peak stress due to a unit vertical force acting at the support cross-section at the vertical position η (Fig. 1.10 a).

As an example, the influence lines of the peak stresses in the flanges of a girder with cross-section having dimensions as indicated in Fig. 1.2 are shown in Figs. 1.10 b,c. The full curves indicate the stresses occurring directly at the support cross-section. It is seen there that the application of a load very close to the flange investigated results in a considerable increase in the peak stresses. This fact is also documented in Fig. 1.10 d, where the distribution of longitudinal normal stresses across the flange breadth is plotted for several position η of the force acting. On the other hand, it has been found that this stress increase has only a local character, disappearing very fast as one proceeds further away from the support cross-section. For instance, the influence lines for the peak stress at cross-sections close to the support are shown in Fig 1.10 b,c by dashed lines.

In practice, cases are also encountered where the girder is supported under

(a)

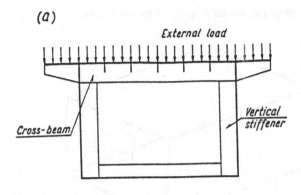

(b)

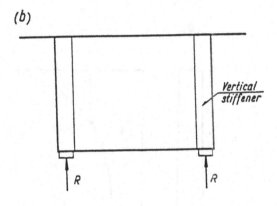

(c)

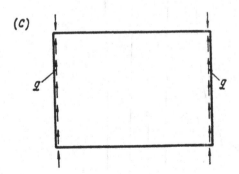

Fig. 1.9. (a) Action of external load on box girder bridges, (b) action of reactions in the support cross-section with vertical stiffeners, (c) shear flow transmitted from the vertical stiffener into the web.

The Main Features of Shear Lag and the Influence of Various Parameters

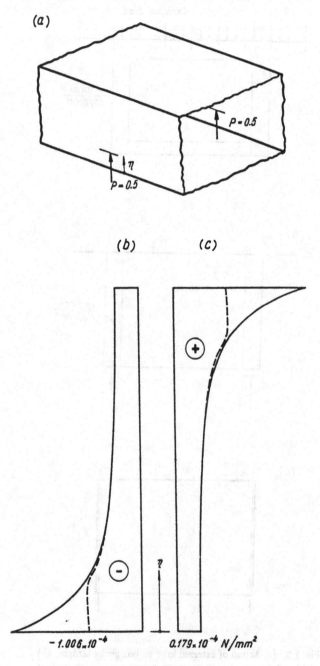

Fig. 1.10.a, b, c

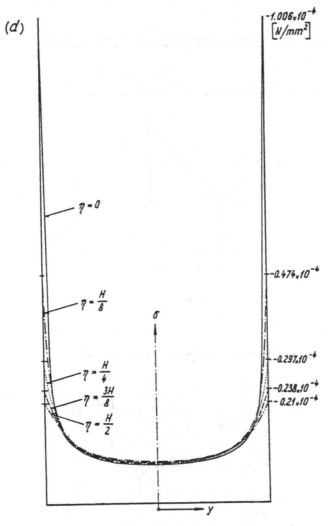

Fig. 1.10. Influence lines of longitudinal normal stress values at the flange edges: (a) construction of the influence lines – positions of a unit loading force, (b) influence line of the value of longitudinal normal stress at the edge of the botton flange, (c) influence line of the value of longitudinal normal stress at the edge of the top flange, (d) distributions of longitudinal normal stresses for various positions of the unit loading force.

diaphragms, which then transfer the reactions into webs (Fig. 1.11). The distribution of the shear flow acting between the diaphragm and the web depends on the positions of the supports. When the support is situated just under the web (Fig. 1.12.a), the shear flow is distributed very non-uniformly, and the distribution of longitudinal stresses across the bottom flange is quite different

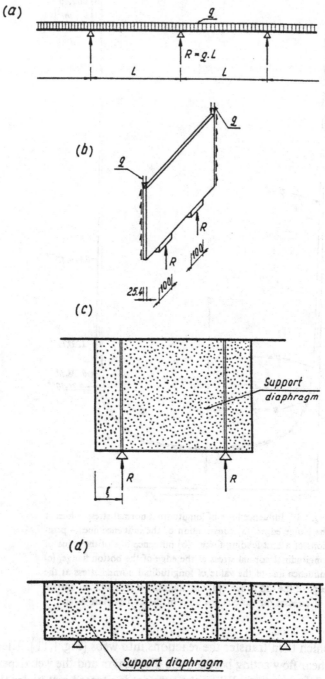

Fig. 1.11. (a) A continuous beam with many spans, (b) action of a
support diaphragm, (c) a support diaphragm in a single-cell box girder,
(d) a diaphragm supporting the central girder.

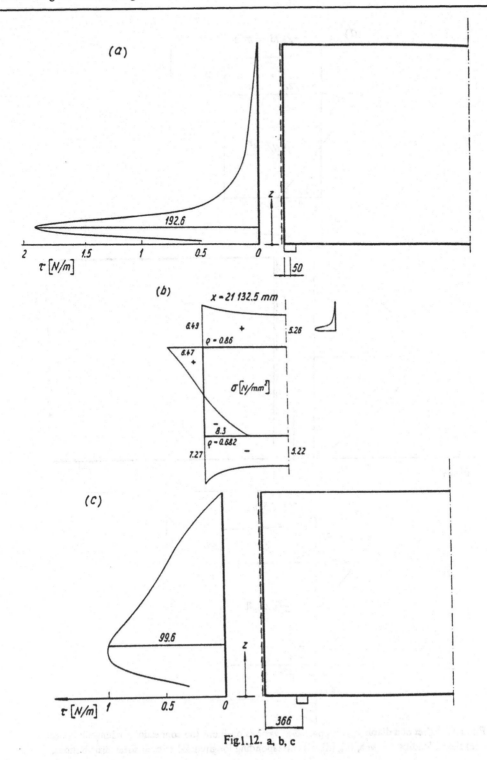

Fig.1.12. a, b, c

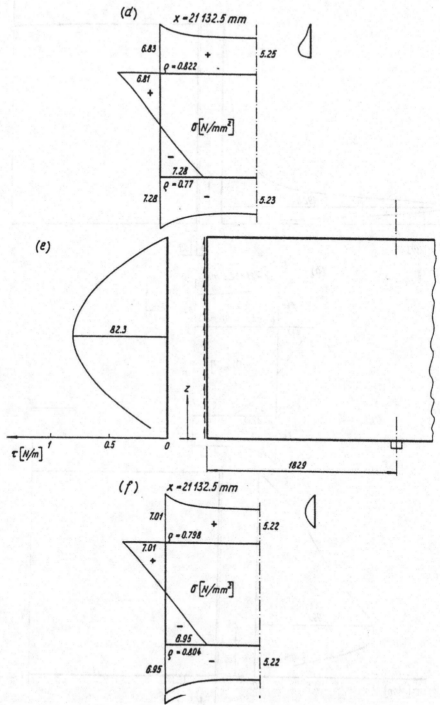

Fig. 1.12. Effect of a diaphragm supported at various positions (no connection to flanges): (a), (c), (e) shears loading the web, (b), (d), (f) corresponding longitudinal normal stress distributions.

(Fig. 1.12b) from that in the top flange. An improvement is achieved when the diaphragm is supported at some distance from the web (Fig. 1.12c,d), and the best state occurs with the support located in the middle of the diaphragm (Fig. 1.12e,f).

These results (Fig. 1.12 are based on a study carried out for a continuous beam with rather long spans. The effects discussed would be much more pronounced with girders having higher width-to-span ratios and/or flanges with longitudinal stiffeners.

1.2.5   Influence of Loads Acting above Longitudinal Stiffeners

When calculating shear lag effects, it is usually assumed that the load acts directly above the webs. This requires a perfect transfer of all loads in the transverse direction (usually through a system of transverse diaphragms). However, cases may be encountered in design practice where short-span bridges provided with rather stiff longitudinal ribs (e.g. railway bridges) are loaded directly above these stiffeners.

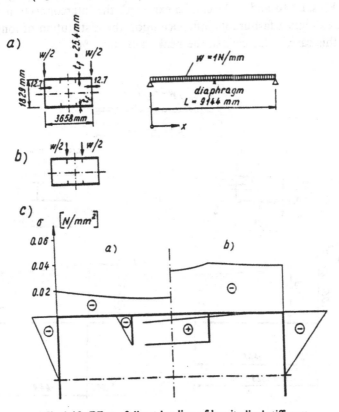

Fig. 1.13. Effect of direct loading of longitudinal stiffeners.

An example of such an arrangement is shown in Fig. 1.13b. Here, a uniformly
loaded girder with central transverse diaphragm is considered. The distribution
of longitudinal stresses at the quarter-span is compared in Fig. 1.13c for a load
acting above the stiffeners (Fig.1.13 b) and the more common case when the load
acts above the webs (Fig. 1.13 a). An entirely different character of the stress-
distribution pattern is apparent. The directly loaded stiffeners behave almost as
independent girders with typical stress distribution along their depths accompanied by their own shear lag effects.

1.2.6   Influence of Overhanging Flanges

It is common practice in the design of steel box girder bridges that the webs
and flanges are interconnected as shown in Fig. 1.14c. The influence of this
arrangement upon the distribution of longitudinal stresses in comparison with
the distribution corresponding to the case usually considered (Fig. 1.14b) is
shown in Figs. 1.14d and e. As can be expected, the interconnection shown in
Fig. 1.14c exhibits a favourable influence upon the distrution of longitudinal
stress (in this case, a decrease in the peak value by 9 %).

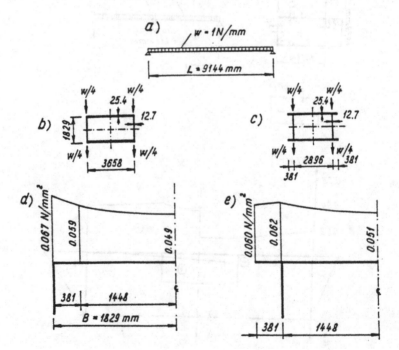

Fig. 1.14. Effect of the arrangement of interconnection of flanges to webs.

1.2.7   Shear Lag in Box Girders with a Flange Stiffened by a Concrete Layer

The previous Sections have shown the shear lag effect to be particularly significant in three specific circumstances:

a) in the regions immediately adjacent to applied concentrated loads,

b) in flange plates that are wide in relation to their spans,

c) in flanges that are longitudinally stiffened.

In the context of simply supported box girders, these three circumstances are rarely coincident. Normally, the upper flange which is stiffened against buckling in compression is protected from the direct incidence of concentrated loading by the surfacing of the bridge deck. The applied wheel loading is thereby dispersed through the thickness of the surfacing and appears as a patch loading on the stiffened flange. Furthermore, the girder proportions are normally such that the span/width ratio of the flange exceeds the value five, below which the shear lag effect becomes particularly significant.

However, in the case of a continuous bridge girder, as in Fig 1.15a, the three circumstances listed above can coincide to produce an exceptionally high shear lag effect, particularly in the lower flange in the region of the intermediate support. The bridge bearings at such points are often positioned underneath the girder webs and bear almost directly on to the girder; they therefore act almost as concentrated loads. Furthermore, the points of contraflexure may well be quite close to the intermediate support so that effective "span" l_s (Fig. 1.15a) of the girder is short, and the resulting flange span/width ratio may well then fall in the critical range below 5. Finally, within such a region of hogging moment, the lower flange may be heavily stiffened against buckling under longitudinal compression.

(a)

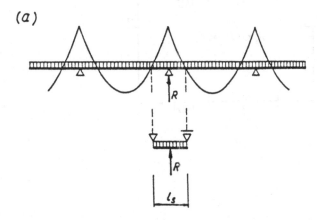

Fig.1.15. a

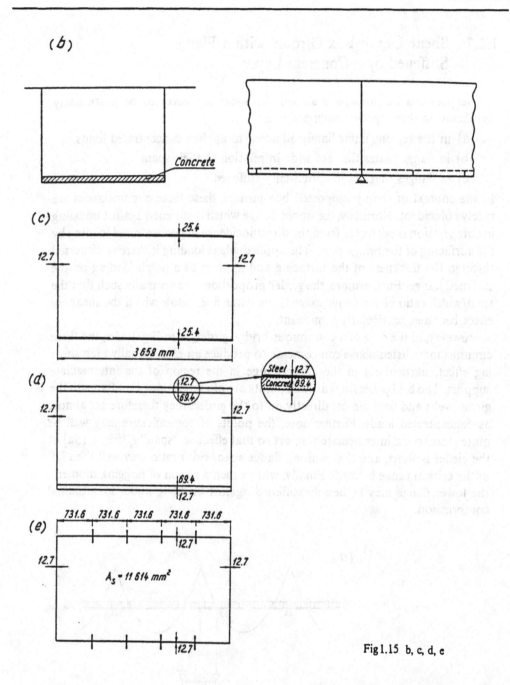

Fig 1.15 b, c, d, e

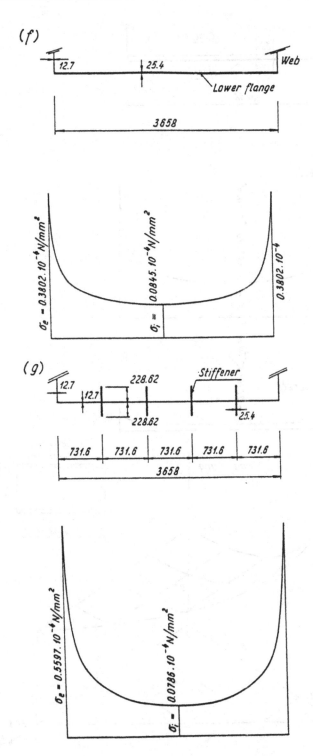

Fig. 1.15 f, g

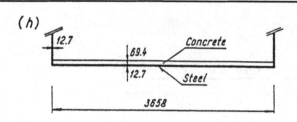

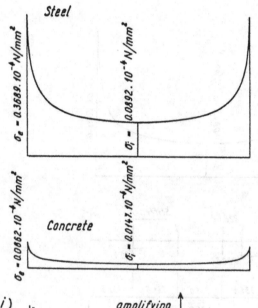

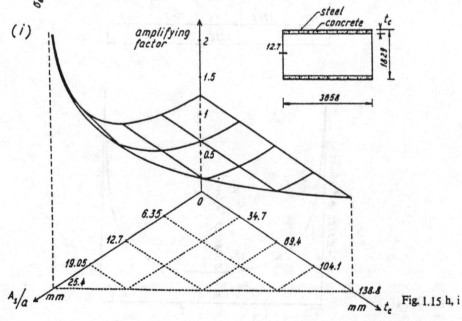

Fig. 1.15 h, i

Little can be done to control the first two of these circumstances. However, a substantial reduction of the third effect, i.e. the shear lag developed as a result of longitudinal stiffening, is possible.

It has been shown that the extent of shear lag within a flange plate is dependent on the ratio between the axial stiffness and the shear stiffness of the plate. The introduction of longitudinal stiffeners increases the axial stiffness without changing the shear stiffness so that there is a consequent increase in shear lag. Stiffeners are, of course, introduced to increase the resistance of the compressed flange to buckling. It would be far more advantageous, from the point of view of shear lag, if the thickness of the plate were increased to make it less prone to buckling, so that stiffeners could be omitted. By increasing the plate thickness, the shear stiffness is increased as well as the axial stiffness, so that the ratio between them, and consequently the extent of shear lag, is hardly altered. It is therefore far preferable to increase the flange plate thickness to resist buckling than to stiffen the plate in situations that are particularly sensitive to shear lag. This may well be possible when a new girder is being designed. However, when it becomes necessary to strengthen an existing girder, possibly to accommodate an increase in bridge loading, then an increase in overall plate thickness is difficult to achieve.

It may be proposed, in such a situation, to stiffen the flange plate with a layer of concrete that is made to act compositely with the steel plate. The necessary composite action can be achieved by means of shear studs welded to the steel plate. The concrete is particularly easy to place on to the bottom flange of the box at an intermediate support, as in Fig. 1.15 b. This is, of course, the area of particularly severe shear lag. It should also be noted that since this additional concrete is placed close to the support, the additional self-weight created in the girder is of little consequence.

The additional layer of concrete does not increase the shear lag effect since it increases the shear stiffness of the flange as well as its axial stiffness. This may be demonstrated by a study whose results are presented in Figs. 1.15 f,g,h. Three box girders with cross-sections shown in Figs. 1.15 c,d,e were analysed. Each girder was simply supported over a span of 9144 mm and the width was 3658 mm to give a span/width ratio of 2.5. Vertical loads of equal magnitude were applied to each web at mid-span, the loads being distributed over lengths of 100 mm to simulate a

◀

Fig. 1.15. (a) A support region of a continuous beam, (b) a box girder bridge stiffened by a concrete layer, (c) flanges without stiffeners, (d) flanges with concrete layers, (e) flanges with steel stiffeners, (f) distribution of longitudinal normal stresses across the flange width for the case without any stiffeners, (g) distribution for the case with steel stiffeners, (h) distribution for the case with a concrete layer, (i) influence of stiffeners and a concrete layer upon longitudinal normal stresses.

practical „concentrated" loading condition. However no vertical stiffening of the web at the point if load application was considered in this study. Each girder was analysed by the folded plate method, which is of proven accuracy and has been fully described in [1.1], [1.6]. Figure 1.15c shows a girder with unstiffened steel flange plates of 25.4 mm thickness. The predicted distribution of longitudinal stresses across the flange at mid-span is plotted in Fig.1.15f and clearly shows the importance of shear lag even in an unstiffened plate.

Figure 1.15e shows a box of similar overall dimensions but with the flange thickness reduced to 12.7 mm and with four longitudinal stiffeners welded to each flange. The stiffener dimensions have been chosen so that the cross-sectional area of the flange and approximately also the overall second moment of area of the cross-section, are the same as those for the first girder; the stress from simple beam theory would thus remain unchanged. However, the actual stress resulting from shear lag reaches exceptionally high values – the severity of the shear lag effect is clearly shown by the non-uniformity of the stress distribution plotted in Fig. 1.15g. Figure 1.15d shows the composite cross-section, again taking a flange plate of 12.7 mm thickness as in Fig. 1.15e, but in this case, with the flange stiffening provided by a 69.4 mm thick concrete layer, acting compositely with the steel plate. The concrete thickness has been chosen to preserve the same equivalent flange area and second moment of area of cross-section as for the two girders in Figs. 1.15c and e. The stress distribution plotted in Fig. 1.15h for the composite flange shows the shear lag effect to be much less severe than in the stiffened flange. The distribution for the composite flange shows characteristics that are very similar to those for the thick steel flange of Fig. 1.15f. Indeed, the values show that the shear lag effect in the composite flange is even less than in the thick steel flange. This slight improvement is due to the fact that concrete has a lower value of Poisson's ratio than steel. Thus, the ratio between the elastic and the shearing moduli and, consequently, the ratio between the axial and shearing stiffnesses, is lower for concrete than for steel; this results in a slight reduction in the shear lag effect.

This last point is of little relevance. The significant point arising from Fig. 1.15h is the marked stress reduction achieved by stiffening the flange with the composite concrete slab instead of with the usual stiffening ribs. Of course, it must be appreciated that, as with any strengthening system that is added to an existing girder, the stress distribution arising from the self-weight of the original girder remains unaffected. However, the strengthening system considerably reduces the stresses from live loads.

In order to illustrate the interrelations between the influences of stiffeners and the concrete layer clearly, another parametric study was carried out.

A box girder with a span $L = 9144$ mm, under a uniformly distributed loading, was analysed. The cross-section, created as a combination of Figs. 1.15c, d and e, is provided with both a concrete layer and steel stiffeners, Fig.

1.15 i. In the s*udy, the dimensions vary so that the modified flange thickness $\bar{t}$ remains unchanged

$$\bar{t} = t + \frac{A_s}{a} + t_c \frac{E_c}{E} = \text{constant} = 25.4 \text{ mm} .$$

Assuming $E/E_c = 5.46$, all three parameters were varied in the following ranges:

(i) the thickness t of the steel flange:

range 0 . . . . 25.4 mm ,

(ii) the thickness t_c of the concrete layer:

range 0 . . . . 69.4 mm ,

(iii) the smeared stiffener area A_s/a:

range 0 . . . . 25.4 mm .

The analysis for this parametric study was carried out by the simple harmonic method presented in   [1.1] . The resulting longitudinal normal stresses at the edge of the flange at mid-span of the girder are expressed in Fig. 1.15 i as an amplifying factor for the stress predicted by simple beam theory. The following behaviour appears:

a) When A_s/a approaches 25.4 mm, if $t_c = 0$, then $t \to 0$. This limit case means that the flange degenerates into a system of isolated stiffeners without any shear capacity. The amplifying factor thus tends to infinity.

b) On the other hand, the substitution for steel stiffeners of a concrete fayer has a benefical influence upon the stress distribution. This phenomenon is most pronounced for the case without any stiffeners $(A_s/a = 0 -$ see the right-hand vertical coordinate plane in Fig. 1.15 i).

In the practical design of structures the exact folded plate theory, or the simple method of harmonic analysis [1.1] intended for use as a fast design tool and available in the form of a program for a personal computer, can be recommended for the analysis of shear lag effects in box girders whose flanges are stiffened by a concrete layer.

1.2.8   Increase in Deflections due to Shear Lag

The considerations presented have concentrated mainly on the distribution of longitudinal stresses across the flange width. However, the shear lag manifests itself also in several further effects, for example in the additional flexibility of the girder, which results in an increase in deflections and which alters the overall bending moment and shear force diagrams in statically-indeterminate structures. The shear lag also affects the pattern of distribution of shear stresses in the flange.

The shear lag in flanges and the shear deformations in webs result in a considerable increase in deflections in comparison with those predicted by the elementary theory of bending. This is particularly pronounced in the short side spans of continuous beams. The actual values of support reactions also differ from those predicted by the ordinary analysis. The increase in the end reactions and the decrease in the inner reactions are caused by the additonal flexibility due to the shear lag. This decrease in the beam stiffness is particularly pronounced in the regions of the inner supports, a result of the effect of the reactions acting as point loads. This is also why the support bending moments are of lower values than those given by the elementary theory of bending – this change of the redundant values results in an increase in the mid-span bending moment.

1.3   Curved Box Girders

Curved box girders, encountered in bridge design practice, can be divided into two groups according to their different structural performance.

1.3.1   Horizontally Curved Box Girder Bridges

Various types of such bridge systems are extensively used in the highway network, taking thus advantage of their rigidity and high resistance to torsional moments as well as of their aesthetically attractive appearance.

For the analysis of horizontally curved bridges, various efficient methods are available besides the finite element method, which represents a general tool; among them particularly for box girders of constant cross-section, the finite strip method (see [1.3], [1.4], [1.5]) is a technique ideally suited for the shear lag analysis

It has been found that the non-uniform character of the longitudinal stress distribution due to shear lag is similar to that in straight girders. Figure 1.16 compares the results of the behaviour of (i) a curved and (ii) a straight box girders. The uniformly distributed load is placed above the web segments. Although the intensity of loads acting on the individual webs is different, the resultant load on each web is the same. As can be seen, these stresses tend to increase nearer the inner web. This effect is due to pronounced torsion in the girder, to different lengths and, consequently, to different rigidities of the outer and inner webs.

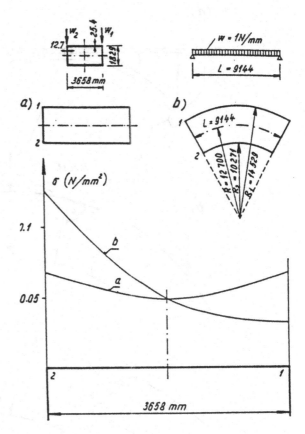

Fig. 1.16. Shear lag in horizontally
curved box girders.

1.3.2 Arches of Box-Shaped Cross-Section

Such structures belong to the most suitable systems for bridges having very long spans. Various shapes of the median line of arch bridges are used in practice in order to minimize the bending effects in the arch. The intention is to accommodate the median line of the arch to the thrust line corresponding to the prevailing loading of the bridge. To be able to take advantage of the use of the finite strip method and to obtain generally valid as well as simple results, the shape of the arch is here approximated by a circular line. This shape is not without practical importance; on the contrary, in many bridges it is better justified than, for example, a parabola of the second order, which represents a thrust line corresponding to uniformly distributed loading.

The structural performance of an arch bridge is characterized by the course of the thrust line (Fig. 1.17 a,b) which (by its distance from the median line of the arch) determines the bending moment diagram (Fig. 1.17 b). The intersec-

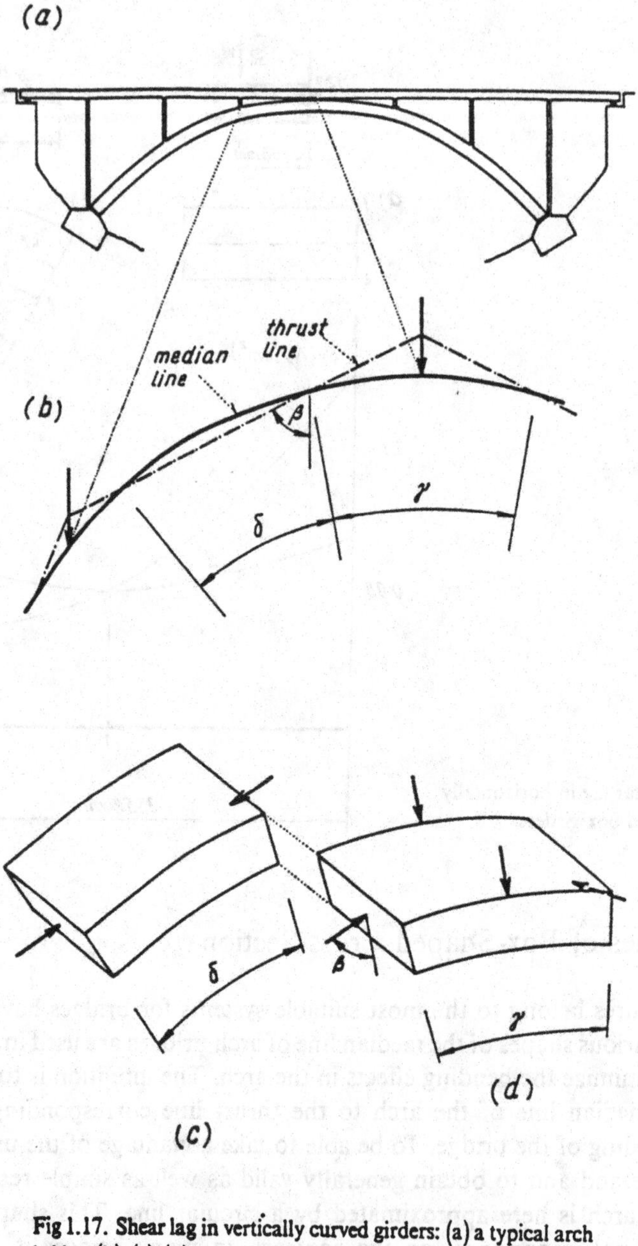

(a)

(b)

median line   *thrust line*

β

δ   γ

(c)

δ   β   γ

(d)

Fig 1.17. Shear lag in vertically curved girders: (a) a typical arch
bridge, (b), (c), (d) secant and intermediate segments of the arch.

tions of the thrust line with the median line yield the locations of points at which the bending moment is zero. The system of loads and reactions acting on each portion of the arch between these inflection points can be evaluated, and the shear lag analysis can then be carried out independently, as an approximation, for each individual portion of the arch. It is seen in Figs. 1.17 b and c that in the arch there are portions of two kinds. The first kind of segment, which can be termed a secant segment, is loaded only by two opposite forces acting at its ends in the direction of the chord of the arch portion. The intermediate segment, on the other hand, is directly loaded by an external force. Both of these kinds of segments, each having its typical structural performance, form self-balanced systems.

It has been found that, as in the case of straight and horizontally curved box beams, the shear lag in arch type structures can play an important role. The non-uniformity of distribution of longitudinal normal stresses across the flange width at the mid-span of the secant segment, shown for the case of an unstiffened flange in Fig. 1.18 a and for a flange with longitudinal stiffeners in Fig. 1.18 b, grows with an increase of the central angle δ, i.e. with an increase of bending action in the more curved segment. Different patterns of stress distributions in the top and bottom flanges are observed as a result of stress relief in the top flange and of a stress increase in the bottom flange due to additional bending caused by the curvature of the segment. It is also seen from Fig. 1.18 a and b that

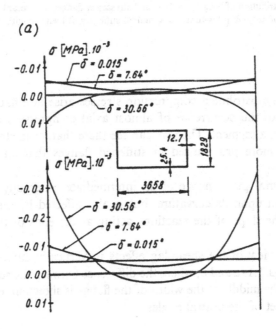

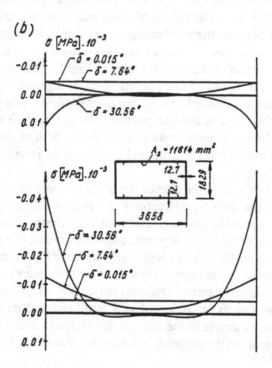

Fig 1.18. Distribution of longitudinal normal stresses in flanges of a secant segment for various values of angle δ: (a) without longitudinal stiffeners, (b) with longitudinal stiffeners.

very uniformly distributed compression stresses arise in a flat segment, which corresponds to the occurrence of almost axial compression along the whole length of such a segment. Then we can see there that, as in straight beams, the shear lag is more pronounced in stiffened flanges than in flanges without stiffening.

The structural performance of an intermediate, externally loaded segment depends again upon its curvature, but is also affected by the angle β, which determines the slope of the reactions acting at the ends of the segment (Fig. 1.17d).

A general view of the shear lag effects in vertically curved box girders is offered in Figs. 1.19a and b, where the difference between the stress values at the edge and in the middle of the width of the flange is shown in terms of the angle β and for a set of the central angles.

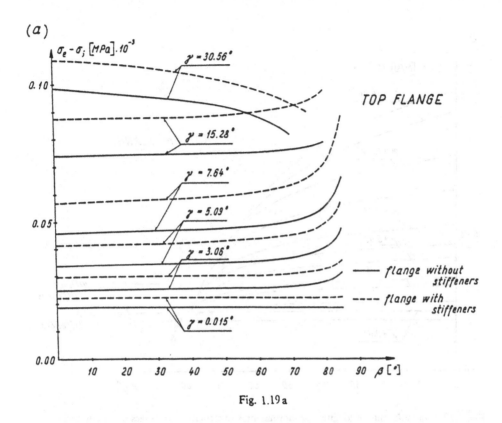

Fig. 1.19 a

1.4 Initial Imperfections in the Shear Lag Behaviour of Wide Flanges

It has been found that in box girders without closely spaced, sufficiently rigid transverse diaphragms the distribution of longitudinal stresses across the width of the flange may be considerably influenced by initial imperfections of the flange sheet (Fig. 1.20). This effect is due not only to a change in the second moment of area of the overall cross-section with imperfections and varying distance of a point at the flange from the cross-sectional neutral axis, but also to the shell action of the non-planar flange sheet loaded along its edges (where the flange is interconnected to the webs) by longitudinal shear flows (Fig. 1.21b).

Various configurations of the initial "dishing" can be encountered in practice; see, for example, Fig.1.20. It can be seen there that the initial curvature may either spread over the whole flange surface, or be only local.

(b)

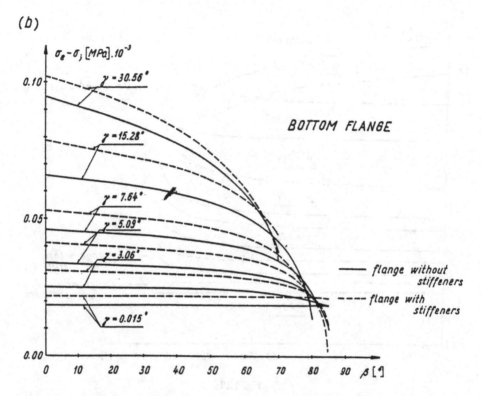

Fig. 1.19. Non-uniformity of longitudinal normal stress distribution in flanges of an intermediate segment in terms of varying angle β: (a) top flange, (b) bottom flange.

The magnitude of initial imperfection can be characterized by its maximum value (amplitude) δ (Fig.1.20). It is obviously of importance whether the inital "dishing" is outward of inward oriented.

1.4.1 Flanges without Longitudinal and Transverse Stiffeners

Let us study the performance of a simply supported girder of the cross-section shown in Fig. 1.22, of length $L = 9\,144$ mm and uniformly loaded by $p = 1$ N/mm. The beam is fitted with cross-beams at its supports, and the flanges exihibit initial imperfections according to Figs. 1.20b–h.

To start with, the influence of a simple sine curve imperfection (Figs. 1.20b and c) with $\delta = \pm b/400, \pm b/300, \pm b/200, \pm b/100$ and $\pm b/50$ is analysed. The

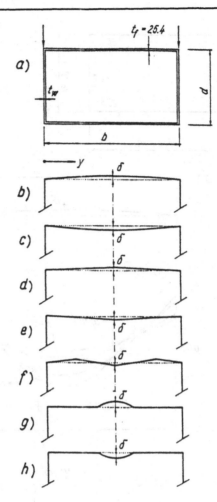

Fig. 1.20. Various types of flange imperfections.

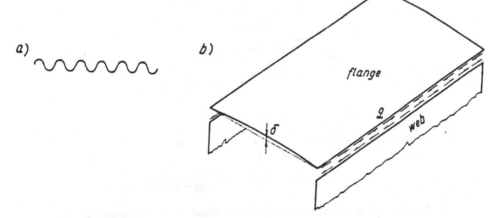

Fig. 1.21. Interaction between flages and webs.

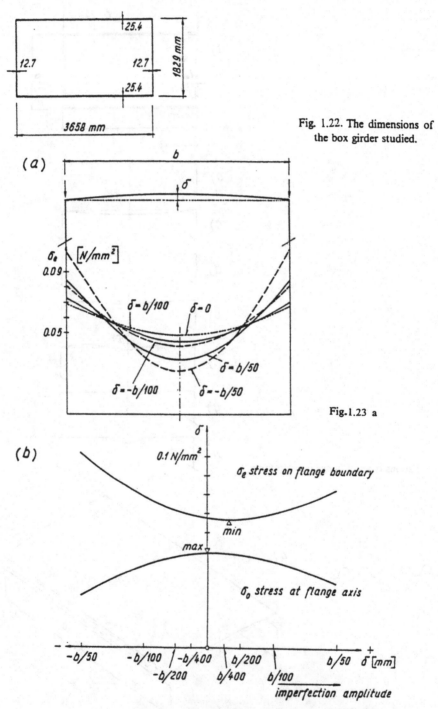

Fig. 1.22. The dimensions of
the box girder studied.

Fig.1.23 a

Fig.1.23. (a) Effect of a sine imperfection – unstiffened flanges, (b) relationship
between the characteristic flange stresses and the imperfection amplitude –
unstiffened flanges.

redistribution of longitudinal normal stresses over the flange breadth at the mid-span of the girder is shown, for several values of imperfection, in Fig.1.23a.

It can be seen there that the effect of initial irregularities is very significant and unfavourable. Thus, in comparison with an ideally plane flange, the stress on the flange boundary increased for $\delta = b/50$ by 22 % when the imperfection was outward-oriented, and by 51 % in the case of inward-oriented "dishing". This difference in behaviour in terms of "dishing" orientation can only partly be explained by the change in moment of inertia discussed above. Much more important is the influence of the longitudinal deflection of an initially curved flange. If the flange curvature is convex, then the additonal longitudinal stresses which occur as a result of the flexure are compressive in the middle of the flange and tensile on its boundaries. This means that their character is opposite to the manifestation of the shear lag phenomenon. If the flange curvature is concave, the effects of the longitudinal flange flexure have opposite signs and, therefore, are added to those of shear lag.

It is of interest to demonstrate the relationship between (i) the stress, σ_e, on the flange boundary and the stress, σ_0, at the flange axis and (ii) the magnitude of imperfection. This is done in Fig. 1.23b. The relationship is very non-linear, which means that the non-uniformity of stresses grows much faster than the magnitude of imperfections. It follows that for small imperfections (say, up to the value of $b/150$) their effect is not, from the practical point of view, significant.

An examination of Fig. 1.23b also indicates that, in line with the aforesaid consideration about the flexure of the flange as a whole, the minimum value of the stress σ_e at the flange longitudinal edge does not correspond to an ideally plane flange, but to a flange plate having an outward-oriented imperfection $\delta \doteq b/300$.

The distribution of longitudinal normal stresses over the width of the flange for some other kinds of imperfections is presented in Fig. 1.24. A very unfavourable effect of initial irregularities is conspicuous particularly for the imperfection studied in Fig.1.24.b, where for larger magnitudes of δ the stress on the flange boundary can attain very high values. It has also been found that the impact of initial irregularities depends rather on their configuration than on their absolute magnitude. Figure 1.25c shows a comparison between the distribution of the longitudinal stresses at mid-span of the box girder considered and having a flange with two kinds of imperfections (as shown in Figs. 1.25a and b) of the same severity $\delta = b/100$. It can be seen that the shape of the imperfect surface of the flange affects the stress distribution pattern more than the absolute severity of the imperfection δ itself. Thus, imperfections as shown in Fig.1.25b have more adverse effects than those shown in Fig.1.25a, although δ is the same. This is in agreement with the well known fact that a decrease in the shear load-carrying capacity occurs in a corrugated sheet (Fig. 1.21a) whose axial load-bearing capacity remains unchanged and results in more pronounced shear lag effects than that of perfectly plane flanges (similar to a flange orthotropy).

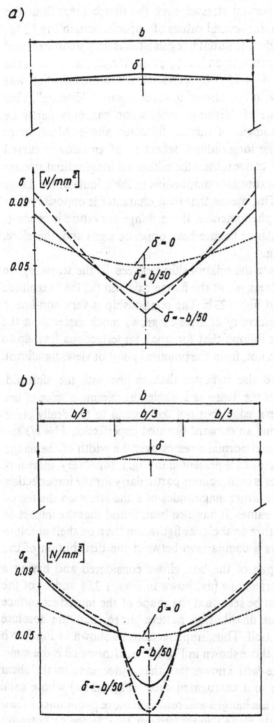

Fig. 1.2-1. Effect of various types of
imperfections – unstiffened flanges: (a)
a roof-type imperfection, (b) a local
imperfection.

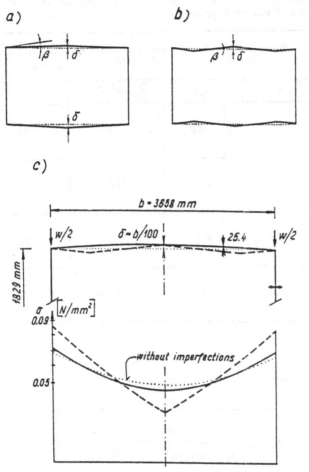

Fig. 1.25. Comparison of the effect of various imperfection types on the distribution of longitudinal normal stresses – unstiffened flanges.

1.4.2   Flanges with Longitudinal but not Transverse Stiffeners

In order to exclude the influence of the eccentricity of single-sided stiffeners (which, as was shown in Section 1.2.3, can affect the character of shear lag), flanges fitted with symmetric double-sided ribs are considered in this study (see Figs. 1.26 a, b and c). Moreover, the total area of the stiffened flange is required to be the same as in the foregoing case of a flange plate without ribs. Therefore, the thickness of the flange sheet is chosen to be one half of the previous thickness, i.e. $t_f = 12.7$ mm.

The results for various types of initial curvature are given in Figs. 1.16 a, b and c. A comparison of the data obtained is also presented in Table 1.1.

Table 1.1. *The Effect of Various Types of Initial Imperfection on Shear Lag in Longitudinally Stiffened Flanges (without Transverse Stiffeners).*

Imperfection	Value of stress σ_e at longitudinal edges		Value of stress σ_e at flange axis	Drop in stress at flange axis with respect to stress at longitudinal edges
	$[N/mm^2 \times 10^{-3}]$	$[\%]$	$[N/mm^2 \times 10^{-3}]$	$[\%]$
$\delta = 0$	-87.67	100	-45.38	48.24
$\delta = b/50$	-97.87	111.7	-38.28	60.90
$\delta = b/50$	-101.42	115.7	-31.62	68.82
$\delta = b/50$	-140.39	160.1	-3.23	97.70

Fig. 1.26.

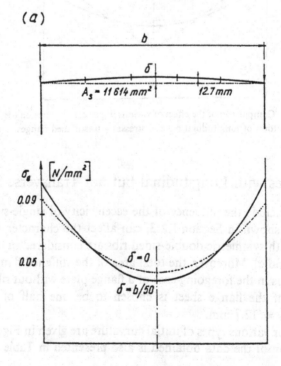

(a)

(b)

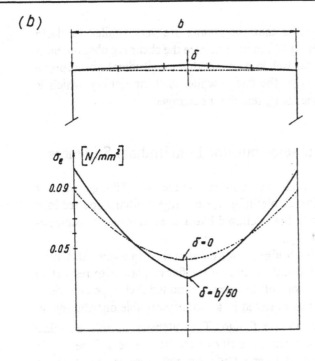

(C)

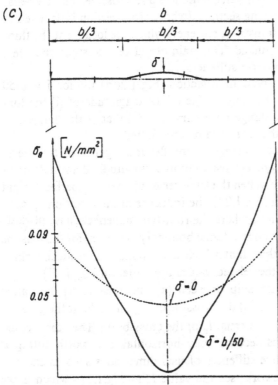

Fig. 1.26. Effect of various types of imperfections – longitudinally stiffened flanges: (a) a sine imperfection, (b) a roof-type imperfection, (c) a local imperfection.

An analysis of the data shows that the results are very similar to those obtained for flange plates without stiffeners. But here the shear lag phenomenon is even more intensive, due to the well known fact that longitudinal ribs improve the axial (longitudinal) rigidity of the flange while its shear rigidity, which is materialized by the flange sheet only, remains unchanged.

1.4.3 Flanges with Transverse but not Longitudinal Stiffeners

It was demonstrated in the previous sections that in the case of flanges without transverse stiffeners, the unfavourable influence of larger initial imperfections was significant. The flange plate then behaved like a shell, this being accompanied by deformation of the flange.

An ordinary steel box girder bridge, however, contains transverse stiffening elements which reduce the relative deflection of the flange plate with respect to the girder webs. The manifestations of the abovementioned shell-type behaviour are thereby diminished, which is reflected in a more favourable distribution of longitudinal normal stresses in the flange. The intensity of this beneficial phenomenon depends on the flexural rigidity of the transverse stiffeners.

To examine the function of transverse stiffeners and cross-beams, the behaviour of a girder as in the foregoing sections (whose cross-section is depicted in Fig. 1.22), but fitted with two transverse stiffeners which are located at the third points of the girder length, is studied. The main objective is to study the effect of the flexural rigidity of transverse stiffeners.

The results related to a roof-type imperfection (Fig. 1.20 d) and for $\delta = b/50$ are given in Fig. 1.27, where the dependence of the magnitude of (i) the longitudinal normal stress on the flange boundary, and (ii) that at the flange axis upon the flexural rigidity of the cross-beams are plotted.

It can be seen there that an increase in the flexural rigidity of transverse stiffeners affects the uniformity of the distribution of longitudinal normal stresses in the flange very favourably. When the transverse stiffeners are perfectly rigid (see the dashed straight line in Fig. 1.27), the influence of imperfections practically vanishes. In the case studied, where the roof-type imperfection is outward-oriented, the value of the stress on the flange boundary becomes (for the reasons discussed above) even lower than that which corresponds to an ideally plane flange without irregularities (the related values are dotted in Fig. 1.27).

The curves giving the relationship between (a) the stress at (i) the flange boundary, (ii) the flange axis and (b) the cross-beam rigidity exhibit the greatest inclination for small moments of inertia, I_t, of the cross-beam. Then the inclination diminishes and the curves become quasi-horizontal. It is worth noting at this juncture that for transverse stiffeners of usal dimensions such as encountered in ordinary steel bridgework (see the value I_t^* in Fig. 1.27, which is the

optimum rigidity of the transverse rib from the point of view of flange buckling
– the corresponding cross-section of the rib is also shown in Fig. 1.27) the effect
of imperfections on the distribution of stresses in the flange is very considerably
reduced.

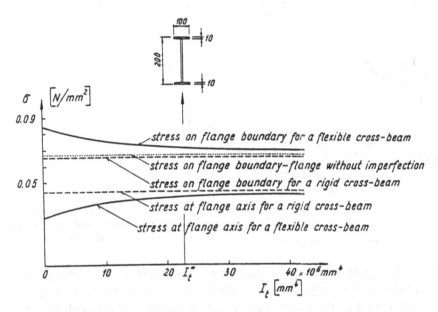

Fig.1.27. Relationship between the characteristic flange stresses and the flexural rigidity of
transverse stiffeners – transversely stiffened flanges, roof-type imperfection.

1.4.4 Flanges with Longitudinal and Transverse Stiffeners

The performance of flanges fitted with both longitudinal and transverse
stiffeners is studied on the same girder and for the same geometrical imperfections as was the case in Section 1.4.2, i.e. in the Section which deals with the shear
lag phenomenon in longitudinally stiffened flanges. However, on top of that, the
girder is stiffened by two cross-beams situated at the third points of the girder
span. The aim is to look into the distribution of longitudinal normal stresses
over the flange width, and to find out how this distribution is influenced by the
flexural rigidity of the cross-beams.

Figure 1.28 shows, as an example, the distribution of longitudinal normal
stresses for a roof-type imperfection and rigid transverse stiffeners, the corresponding curve for an "ideal" flange without irregularities being also plotted
there for the sake of comparison. It can again be seen that sufficiently rigid
transverse stiffeners are able to reduce considerably (or practically eliminate) the
effect of geometrical irregularities. The slight difference between the two curves

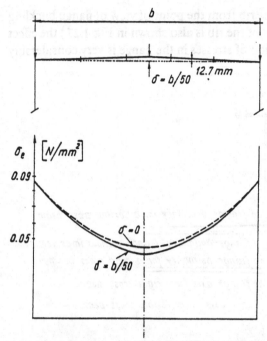

Fig. 1.28. Effect of a roof-type imperfection – longitudinally and transversely stiffened flanges.

shown in the figure is because, in the case of an initially imperfect flange, the
distance between the flange points and the horizontal axis is variable, and the
moment of inertia of the whole cross-section in this case of outward-oriented
"dishing", in comparison with the moment of inertia of an "ideal" cross-section
without imperfections, is a little enlarged.

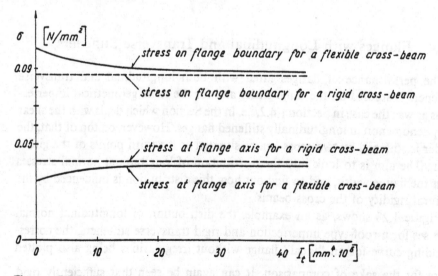

Fig. 1.29. Relationship between the characteristic flange stresses and the flexural rigidity of
transverse stiffeners – longitudinally and transversely stiffened flanges, roof-type imperfection.

Figure 1.29 gives for the same girder and imperfection the relationship between (i) the characteristic values (on the flange boundary and at the flange axis) of longitudinal normal stresses and (ii) the flexural rigidity of transverse stiffeners.

A comparison of the magnitudes of characteristic stresses in a longitudinally stiffened flange, with and without cross-beams, is presented in Table 1.2 .An examination of that table reveals that the presence of transverse stiffeners always brings about a reduction of the peak values of longitudinal normal stresses in the flange; this means that the stress values at the longitudinal flange edges are reduced while those at the flange axis are increased. The beneficial effect of cross-beams is a function of the configuration of the initial flange "dishing".

The results obtained make it possible to formulate more accurately the role of transverse stiffeners on the shear lag behaviour of the wide flanges of steel box girder bridges.

Table 1.2. *The Effect of Various Types of Initial Imperfection on Shear Lag in Longitudinally Stiffened Flanges.*

Imperfection		Girder without cross-beams		Girder with cross-beams	
		$\sigma_x \left[N/mm^2 \times 10^{-3} \right]$	$[\%]$	$\sigma_x \left[N/mm^2 \times 10^{-3} \right]$	$[\%]$
(fig.)	σ_{xe}	-97.87	100	-86.77	88.65
	σ_{xo}	-38.27		-42.91	112.14
(fig.)	σ_{xe}	-101.42	100	-86.85	85.64
	σ_{xo}	-31.65		-41.34	130.60
(fig.)	σ_{xe}	-140.39	100	-108.27	77.12
	σ_{xo}	-3.23		-27.95	865.85

When the flange is perfectly plane (i.e. without initial irregularities) and without longitudinal ribs, or when the flange is plane and these ribs are symmetric with respect to the middle plane of the flange sheet (for instance, double-sided symmetric ribs), the flange is in a plane state of stress, and its transverse deflection is insignificant. In such cases, the flexural rigidity of cross-beams is not able to play any pronounced role. This is also the way the problem is looked at by most practising engineers.

The situation is, however, quite different when the above conditions are not fulfilled, and when the flange tends to deflect. Then the presence of sufficiently rigid transverse stiffeners is very effective, since they are able to hinder the deformation of the girder cross-section, and the effect of imperfections on shear lag can then be disregarded, as is the usual current practice in bridge design.

Hence, the effect of cross-beams and transverse ribs is indirect; by helping to preserve the cross-section shape, they are able to reduce substantially the unfavourable influence of geometrical flange imperfections on shear lag behaviour.

1.5   Box Girders Subjected to Torsion

It is well known that the technical theory of warping torsion provides a linear distribution of longitudinal normal stresses over the widths of individual plate elements, as shown in Fig. 1.30. It can be expected also that in the case of box girders under torsion a redistribution of longitudinal normal stresses occurs in the plate elements of the box girder, similar to that which develops, due to shear lag, in the flanges of box girders during bending.

The study is performed on the same box girders as were used in the above sections. The stress states obtained, at the mid-span and at the quarters of the girders, are plotted in Figs. 1.31–1.34.

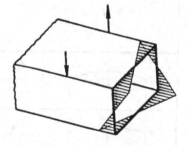

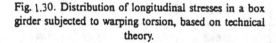

Fig. 1.30. Distribution of longitudinal stresses in a box girder subjected to warping torsion, based on technical theory.

Figure 1.31 presents results for a girder having a span of 9144 mm which is subjected to a concentrated torsional moment (loaded length 100 mm) at the central section. It can be seen there that the distribution of stress in the upper flange at the loaded mid-span section of the girder is clearly curvilinear (see Fig. 1.31b), thereby deviating significantly from the basic assumption of the technical theory. On the other hand, the distribution in the lower flange at the same section is close to a straight line, and at the quarter-points of the girder, which are farther away from the loaded mid-span section, the agreement with the technical theory is very good (Fig. 1.31c).

The results of a study of the effect of uniformly distributed torsional loading on the same girder (having a span of 9144 mm) are given in Fig. 1.32. The distribution of the stresses is almost perfectly rectilinear, indicating that application of the technical theory is fully justified in this case.

For shorter spans, however, considerable deviations from the straight-line assumption occur both for the concentrated twisting moment (Fig. 1.33) and for the uniformly distributed torsional loading (Fig. 1.34). In the former case, the deviations dominate over a substantial part of the girder length.

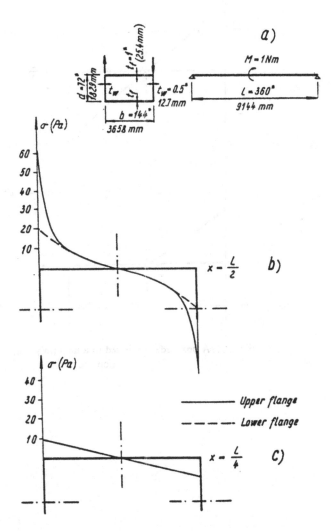

Fig. 1.31. A box girder subjected to a concentrated torsional moment.

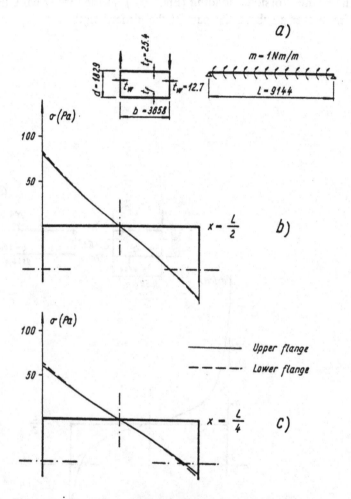

Fig. 1.32. A box girder subjected to a uniformly distributed torsional
moment.

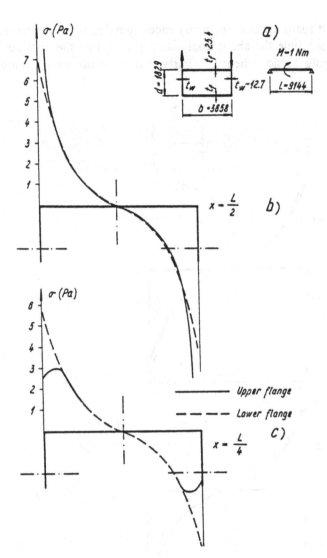

Fig. 1.33. A short box girder subjected to a concentrated torsional moment.

As such short spans are not frequently encountered in ordinary structures, it is possible to conclude the above analysis by stating that the applicability of technical warping torsion theory is fairly wide particularly for continuous loading.

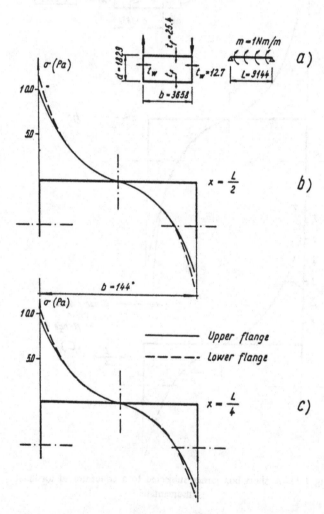

Fig. 1.34. A short box girder subjected to a uniformly distributed torsional moment.

REFERENCES

1.1 GIRKMANN, K.: Flachentragwerke, Springer Verlag, Wien 1959.
1.2 MOFFAT, K., R. and DOWLING, P. J.: Shear lag in steel box girder bridges, Struct. Eng., No. 10, 1975.
1.3 CHEUNG, Y., K.: The finite strip method in the analysis of elastic plates with two opposite simply supported ends, Proc. Inst. Civ. Eng., May 1968.
1.4 MEYER, C.: Analysis and design of curved box girder bridges, SESM Report 70-22, University of California, Berkeley, Dec. 1970.
1.5 KŘÍSTEK, V.:Theory of box girders, J.Wiley and Sons, Chichester, 1979.
1.6 KŘÍSTEK, V. and EVANS, H.R.: A hand calculation of the shear lag effect in unstiffened flanges with closely spaced stiffeners, Civil Engineering for Practising and Design Engineers, Vol. 4, No. 2, February 1985.
1.7 EVANS, H. R. and KŘÍSTEK, V.: A hand calculation of the shear lag effect in stiffened flange plates, Journal of Constructional Steel Research, Vol. 4, No.2, 1984.
1.8 KŘÍSTEK, V. and BAŽANT, Z.P.: Shear lag effect and uncertainty in concrete box girder creep, Journal of Struct. Eng., ASCE, Vol. 113, No.3, March 1987.
1.9 KOVANICOVÁ, Z., KŘÍSTEK, V. and ŠKALOUD, M.: Possibility of reducing shear lag effects (in Czech), Inženýrské stavby no. 4, 1985.
1.10 KŘÍSTEK, V., STUDNIČKA, J.and ŠKALOUD, M.: Shear lag in the wide flanges of steel bridges. Acta technica ČSAV, No. 4, 1981.
1.11 KŘÍSTEK, V., ŠKALOUD, M.: The role of initial imperfections in the shear lag behaviour of wide flanges. Acta technica ČSAV, No.6, 1983.
1.12 KOVANICOVÁ, Z., KŘÍSTEK, V. and ŠKALOUD, M.: Shear lag in the wide flanges of vertically curved box girders, Acta technica ČSAV, No. 4, 1988.
1.13 EVANS, H. R., KŘÍSTEK, V. and ŠKALOUD, M.: Shear lag in box girders with a flange stiffened by a concrete layer. Acta technica ČSAV, No.3,1989.

2. "Breathing" of Slender Webs

It has been shown (see e.g. [2.1], [2.2], [2.3]) that the currently used slender webs and flanges of steel girders exhibit a large post-buckled reserve of strength. To make design economical, this post-critical reserve is taken into account, with there being a pronounced tendency during the recent years to exploit not only the elastic but also the plastic portions of this reserve. This presents no special problems in the case of thin-walled girders used in steel buildings, which are subject to quasi-constant loading. But the situation becomes much more complex when dealing with steel bridges, cranes, crane-runway girders and other types of transport and similar structures, which are subject to loads repeated many times. In such cases stability phenomena interact with cumulative damage and fatigue phenomena. Then, a substantial portion of the plastic post-critical behaviour of the system may be "eroded" by the initiation and propagation of cracks in the webs and flanges of the system, this eventually changing completely the failure mechanism of the structure.

As very little is known in this respect, the importance of research on web "breathing" is now generally appreciated.

The character of "breathing" of the plate elements of steel plate and box girders reflects the character of the repeated loading to which the girder is subjected. Such a girder can be subjected to either (i) a large number (millions) of repeated lower loads, or (ii) a small number (hundreds or thousands) of repeated higher loads, or (iii) a combination of the two. Evidence on all aforesaid kinds of "breathing" is needed.

This first stage of the Prague research on web "breathing", which is described here, below, is intended as a contribution to the solution of problems of the above group (ii).

2.1 "Breathing" of webs subjected to repeated partial edge loading

The aim of the first part of the Prague research into web "breathing" was to look into the effect of the repeated character of patch loading on the ultimate limit state of steel plate girder webs without and with longitudinal ribs. This campaign of tests was conducted jointly by K. Januš, I. Kutmanová and the author of this chapter.

2.1.1 Test girders and test set-up

Eighty test girders were subjected to repeated patch laoding, their webs being fitted with a longitudinal stiffener positioned at (i) one tenth, (ii) one fifth of the web depth. For the sake of comparison, a number of girders without longituditinal stiffening were tested, too. The test girders had the same dimensions, and were fabricated from the same material, as those used in the preceding constant patch loading tests, this being so as to enable the invgestigators to compare the results (ultimate loads, onset-of-yielding loads, etc.) of both respective experimental series.

In all experiments, the load strengths, s_0 , was constant, $s_0 = a / 10$. The general details of the test girders can be seen in Fig. 2.1. Their main characteristics are listed in Tables 2.1 a-c.

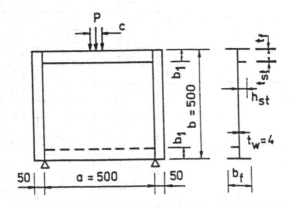

Fig. 2.1: General details of test girders subject to repeated patch loading

The variable repeated loading was materialized by means of a 1000kN AMSLER pulsateor, the frequency of loading cycles being of 3.75 Hz. The following quantities were registered during the tests:

(i) the initial imperfections of the experimental girders,

(ii) the values of deflections and strains at a number of selected places,

(iii) the initiation and prpagation of cracks, and

(iv) the acoustic emission in the web in the neigbourhood of the applied load.

The load P cycled between (i) zero and (ii) a value P_{max} , which in turn was varied between (a) the constant loading ultimate load P^0_u and (b) the onsetof surface-yieldingload detected also in the related constant loading test. Thus, under the above loading, the webs of the girders tested behaved in the elasto-plastic range and, consequently, their performance was expected to be governed by low-cycle fatigue. Therefore, the basic number of loading cycles was chosen so as to be equal to 5×10^4 cycles no failure (whether through initiation of cracks or through excessive plastic buckling of the girder web) occurred, the experiment was continued under a higher load level. If, however, a crack appeared in a certain loading cycle, the experiment went on, under the same (i.e., unchanged) load level, as long as the load-carrying capacity of the test girder was not exhausted.

Table 2.1a: Geometrical and material characteristics of the test girders subject to repeated patch loading - the first stage of experiments

Web		Flange		Longitudinal stiffener			
t_v (mm)	$R_{y,v}$ (MPa)	b_f/t_f (mm/mm)	$R_{y,f}$ (MPa)	b_i/b (-)	h_{st}/t_{st} (mm/mm)	$R_{y,st}$ (MPa)	Notation of test girder
4	303.9	120/8	277.5	0.1	22/8	-	´TG 341-1, TG 341-1´
					20/8	-	TG 341-2, TG 341-2´
					25/8	-	TG 341-3, TG 341-3´
					30/8	290.3	TG 341-4, TG 341-4´
					35/8	271.6	TG 341-5, TG 341-5´
					40/8	275.3	TG 341-6
				0.0			TG 341-6´
		120/20	244.4	0.1	16/8	-	TG 342-1, TG 342-1´
					20/8	-	TG 342-2, TG 342-2´
					25/8	-	TG 342-3, TG 342-3´
					30/8	290.3	TG 342-4,
					35/8	271.6	TG 342-5
					40/8	275.3	TG 341-6
				0.0			TG 342-5´, TG 341-6´

Table 2.1b: Geometrical and material characteristics of the test girders - the second stage of experiments

Web		Flange		Longitudinal stiffener			
t_v (mm)	$R_{y,v}$ (MPa)	b_f/t_f (mm/mm)	$R_{y,f}$ (MPa)	b_i/b (-)	h_{st}/t_{st} (mm/mm)	$R_{y,st}$ (MPa)	Notation of test girder
4	303.9	120/8	277.5	0.0			TG 441-0/1, TG 441-0/1´
							TG 441-0/2, TG 441-0/2´
				0.2	32/8	268.4	TG 441-1/1, TG 441-1/1´
							TG 441-1/2, TG 441-1/2´
					45/8	268.4	TG 441-2/1, TG 441-2/1´
							TG 441-2/2, TG 441-2/2´
		120/20	224.4	0.0			TG 442-0/1, TG 442-0/1´
							TG 442-0/2, TG 442-0/2´
				0.2	32/8	268.4	TG 442-1/1, TG 442-1/1´
							TG 442-1/2, TG 442-1/2´
					45/8	268.4	TG 442-2/1, TG 442-2/1´
							TG 442-2/2, TG 442-2/2´

Table 2.1c: Geometrical and material characteristics of the test girders - the third stage of experiments

Web		Flange		Longitudinal stiffener			Notation of test girder
t_v (mm)	$R_{y,v}$ (MPa)	b_f/t_f (mm/mm)	$R_{y,f}$ (MPa)	h_{st}/t_{st} (mm/mm)	$R_{y,st}$ (MPa)	b_i/b (-)	
4	328.2	120/8	277.5			0.0	TG 541-0/1, TG 541-0/1'
							TG 541-0/2, TG 541-0/2'
				35/8	268.4	0.1	TG 541-1/1, TG 541-1/1'
							TG 541-1/2, TG 541-1/2'
							TG 541-1/3, TG 541-1/3'
						0.2	TG 541-2/1, TG 541-2/1'
							TG 541-2/2, TG 541-2/2'
							TG 541-2/3, TG 541-2/3'
		120/20	244.4			0.0	TG 542-0/1, TG 542-0/1'
							TG 542-0/2, TG 542-0/2'
				35/8	268.4	0.1	TG 542-1/1, TG 542-1/1'
							TG 542-1/2, TG 542-1/2'
							TG 542-1/3, TG 542-1/3'
						0.2	TG 542-2/1, TG 542-2/1'
							TG 542-2/2, TG 542-2/2'
							TG 542-2/3, TG 542-2/3'

During the analysis of the measurements obtained, the experimental test panels of all series were divided into groups: first, according to the position of the longitudinal stiffener ($b_1 = b/10$, $b_1 = b/5$, $b_1 = 0$ - i.e. no stiffener), and then in each of the groups according to flange rigidity (i.e., thin loaded gflange - thick loaded flange). This reflected our experience from the previous constant loading tests, which had demonstrated that stiffener location and flange size were the two main factors that influenced the behaviour of webs in the most significant way.

2.1.2 Failure mechanisms of the test girders

One of the most important questions to which the experiments were expected to give a reply was to what extent the failure mechanisms of girders subject to repeated patch loading were different from those of girders under the action of a constant load.

An examination of the data obtained revealed that in both cases the failure mechanism usually consisted of a set of three platic hinges in the loaded flange and a segmental line plastic hinge in the adjacent zone of the web sheet, which in the case of thin flanges emanated from the two outer flange hinges (then the length of the chord of the segmental line hinge was about one third of the web length) and in the case of thick flanges (when no flange hinges developed) from the web corners (then, of course, the length of the chord of the segmental line hinge equated the web panel length). In addition, during the repeated loading tests, cracks appeared in the web sheet, either in the zone of the (subsequent) segmental line plastic hinge or close to the weld connecting the web sheet to the loaded flange. In most cases, the cracks occurred in the course of the last loading step. In some cases, they developed as late as the failure of the girder, and in others they did not appear at all.

To illustrate this phenomenon, the behaviour of test girder TG 442-1/1 (having a thick flange and $b_1/b = 0.2$), which was first tested under a constant load and then - in an upside down position (as TG 422-1/1') - under a repeated one, is discussed here, below.

The loading of the girder proceeded in four steps. Pending the first three ones (i.e., $P_{max} = 0.8 P^0_u$, $P_{2max} =).85 P^0_u$ and $P_{3max} = 0.9P^0_u$, P^0_u being the ultimate load obtained after 5×10^4 loading cycles. As late as the fourth loading step, under a load of $P_{4max} = 0.95P^0_u$, a crack appeared in the vicinity of the weld (connecting the web sheet with the loaded flange) after 3.76×10^4 cycles. This crack was first detected on the front side of the web and a little later, after 4.89×10^4 cycles, aslo on the rear side. The collapse mechanism developed much later, viz. as late as the 12.65×10^4-th loading cycle of this fourth loading step.

The performance of girder TG 441 - 2/1 (having flexible flanges and $b_1 / b = 0.2$), which was again first tested under constant loading then, as TG 441 - 2/1', in an upside-down position, under a variable fluctuating load, was similar. In the latter case, a crack was detected near the weld as early as the first loading step, i.e. under $P_{1max} = 0.85P^0_u$, after 4.05×10^4 loading cycles. Also this crack first appeared on the front side of the web, but after 4.50×10^4 cycles was also detected on the rear side. The "life" of the girder was completely exhausted after 6.2×10^4 loading cycles.

2.1.3 Initiation and propagation of cracks

The initiation and propagation of cracks, a phenomenon very characteristic of the peformance of webs under repeated loading, was very carefully studied in the course of all the tests. In all experiments, this was done visually with the aid of a magnifying glass; but in the third series of tests, i.e. in the case of 12 test girders, acoustic signal emission checks were emplayed to verify visual observations. It was concluded that, in comparison with the visual observations, the acoustic signal emissionapproach gave more accurate results. For example, this method was able to detect the initiation of cracks 4500-9000 cycles (i.e. 20-40 minutes) earlier. This means that, within the regime of 5×10^4 repeated cycles, the difference was of 9-18%.

Some results of acoustic signal emission checks on girder TG 542-2/2 are given in Fig. 2.2. The curve indicating the relationship between the sum of counts N_c and the number of laoding cycles, transformed into time, well demonstrates an increase in load (point A) and the initiation of a crack and its propagation. The initiation of the crack is associated with the point of contraflexure (point B) of the curve. Point C denotes that momentat which the cracks were detected by visual inspection checks. The inaccuracy of visual inspection is obvious.

Also other pieces of information were used to verify the initiation of cracks. For instance, it was observed that this intiation was accompanied by a sudden change in vertical strains in the respective zones (Fig. 2.3) of the web and, at least with some test girders, also by a fast enlargement of web deflections. Fig. 2.3 again testifies to visual inspection checks being able to detect the initiation of a crack only with a certain delay.

Two other general conclusions may be drawn from the evidence obtained: (i) the propagation of cracks was not uniform and frequently proceeded with more or less long interruptions (see Fig. 2.4, where l denotes the length of crack and N again, the number of loading cycles), (ii) the appearance of a crack by far did not herald the end of the useful life of the girder concerned. The girder was thereafter able to sustain a good many more loading cycles before its load-bearing capacity was exhausted.

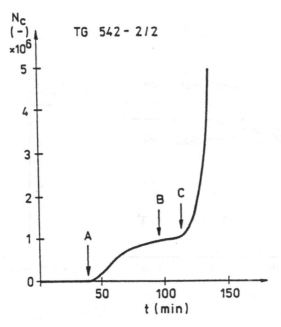

Fig. 2.2: Initiation of crack: comparison of visual inspection with acoustic signal emission checks

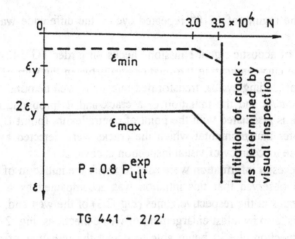

Fig. 2.3: Sudden change in strains indicating the initiation of a crack

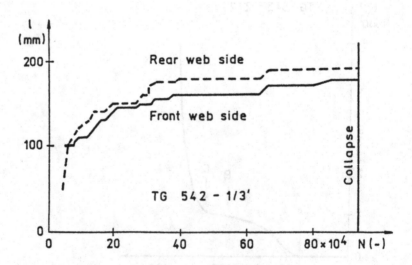

Fig. 2.4: Interruptions in the propagation of cracks

2.1.4 Ultimate limit state of webs subject to a repeated partial edge load

The main test results are summarized in the bar charts in Fig. 2.5. They are plotted in terms of (i) stiffener position and (ii) flange size (the left-hand bar being related to girders with a thin flange and the right-hand one to girders with a thick flange).

The following quantities are given there:

(a) The onset of surface-yielding loads, P_{fat}, determined as the maximum load values under which no appearance of cracks (or of other kinds of failure) was detected (black plus white portions of the bars) after 5×10^4 load cycles.

Both aforesaid quantities are related to P^0_u, i.e. to the ultimate loads obtained in the foregoing constant loading tests.

An examination of the bar charts plotted shows that even in the case of webs "breathing" under the action of a repeated partial edge load a certain (and quite significant) degree of plastification in the webs does not jeopardise the safety of the girders. On the other hand, it can also be seen in Fig. 2.5 that fatigue limit laods are by 10 - 35 % lower than the corresponding constant loading ultimate loads; this testifies to the substantiail influence of "breathing" on the ultimate load performance of webs under the action of a repeated partial edge load. A substantial

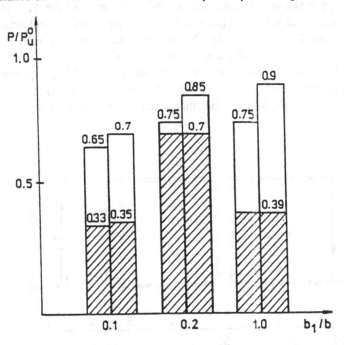

Fig. 2.5: The onset-of-surface-yielding loads and the fatigue limit loads of the test girders subject to repeated patch loading

beneficial influence of flange thickness on fatigue limit loads also follows from the figure. The "erosion" of the plastic portion of the post-critical reserve of strength due to web "breathing" is therefore less pronounced in the case of girders having thick flanges.

2.2 Breathing of webs subject to repeated shear

The aim of the second part of the Prague research on web "breathing", which has been under way for some time, is to study the "breathing" of webs in (predominant) shear. This campaign of tests is being conducted by M. Drdácký, M, Zörnerová and the current author, all of them in the Institute of Theoretical and Applied Mechanics within the Czech Academy of Sciences in Prague.

2.2.1 Test girders and test set-up

Seventeen test girders, all under the action of repeated combined shear and bending, with the effect of shear predominating, were tested during the first stage of the research; and further experiments of the same (or similar) kind are now being conducted.
The general details of the tets girders are seen in Fig. 2.6 a and b.

Figure 2.6 a shows the configuration and dimensions of the pilot girder PBTG, which was tested in the first place. Fig. 2.6. b presents the general details of the main test series BTG 1 - BTG 16. IN this series the diemnsions of the webs were the same, but two kinds of flanges, thick and thin, were used, since it was well known from constant loading tests on shear girders (see [2.3],[2.4]) that, for given aspect and depth-to-thickness ratios of web sheet panels, it was in particular flange size that substantially affected the ultimate load behaviour of such girders.

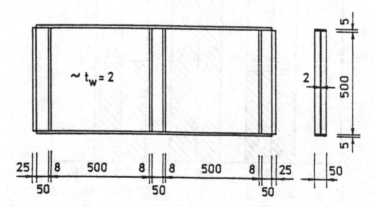

Fig. 2.6a: General details of test girder PBTG

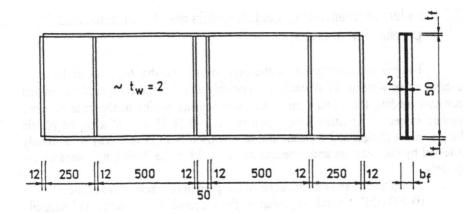

Test Girder	b_f (mm)	t_f (mm)
BTG 1, 2, 6, 8, 9 - 12	50	5
BTG 3, 4, 5, 7, 13 - 16	60	10

Fig. 2.6b: General details of test girders BTG 1 - 16

The material was steel of the same grade as in the case of repeated patch loading tests, viz. Czech Mild Steel 37. In addition, the test set-up and the apparatus employed in the experiments were similar to those used in the aforementioned repeated patch loading tests. The main difference was that the experiments were performend in a 250 kN MTS pulsator, the frequency of loading cycles being of 3 Hz. Obviously, the load was placed above the central transverse stiffener, so that the web panels were subjected to combined shear and bending, with the effect of shear highly predominating.

The loading regime, too, was identical though only during the first stage of the repeated shear loading tests. In the second stage , whose objective was to determine S/N curves for webs subject to repeated shear, each test girder was exposed - from the beginning to the end of the test - to the same loading level. The basic number of loading cycles being again equal to 5×10^4 . If, however, a crack was detected in a certain loading cycle, the experiment went on under the same load level as long as the collapse of the girder was not reached.

It is perhaps worth mentioning at this juncture that a number of new techniques (for example, the method of liquid crystals, thermovision and a laser vibration pattern imager) were used in an attempt to perfect the analysis of web "breathing" and crack growth. However, the main role in crack detection was again played by visual inspection, aided by a magnifying glass and a microscope.

2.2.2 Failure mechanisms of the test girders and the initiation and propagation of craks

The principle conclusion of the experiments was that the ultimate loads of shear girders were in most cases considerably reduced (if compared to related constant loading tests) and their failure mechanisms (which in the case of plate girders subject to constant shear consists - see [2.3], [2.4] - of a set of plastic hinges in the flanges and a plastic diagonal tension band in the web) substantially affected by the initiation and propagation of cracks in the "breathing" webs of the girders.

The initiation and the character of the growth of cracks was as follows:

(i) EITHER a crack appeared in the diagonal tension band and speedily advanced (simultaneously growing in both senses) in a direction more or less perpendicular to that of the tension band (Fig. 2.7) .

This situation, however, happened only once; namely, with girder PBTG; which was tested as a "pilot" girder before the beginning of the main experimental series. It is worth mentioning in this context that PBTG was not a "virgin" girder, since it had already been used in a precesing test, even though not under a high load. In addition, despite the fact that it did not exhibit any significant plastic deformation, it was in the nature of things that a certain cumulative damage had already occurred in the girder before the "pilot breathing" experiment was carried out, with this probably affecting the initiation of the crack in the web.

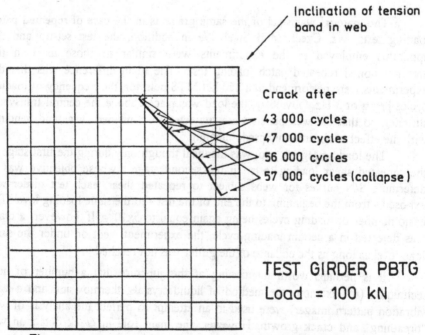

Inclination of tension
band in web

Crack after

43 000 cycles
47 000 cycles
56 000 cycles
57 000 cycles (collapse)

TEST GIRDER PBTG
Load = 100 kN

Fig. 2.7: Advance of a crack in the web of test girder PBTG

(ii) OR a crack started near the inner transverse stiffener, namely, close to that portion of the stiffener into which the diagonal tension band in the web (typical of the post-buckled performance of webs in shear - see [2.3], [2.4],) is anchored, then propagated, fairly slowly, along this stiffener. Thereafter, usually after several thousands of loading cycles, the cracks turned inside (advancing more or less perpendicularly to the diagonal tension band) the web sheet (see Fig. 2.8).

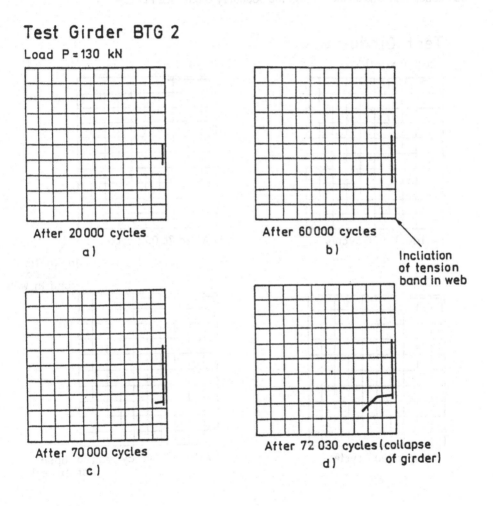

Fig. 2.8: Advance of a crack in the web of test girder BTG 2

(iii) OR a crack initiated (Fig. 2.9) in the close vicinity of the fillet weld joining the upper flange with the web sheet, then propagated (at first in both directions) along the weld so that on one side it reached the adjacent web corner. Then it turned down there, in a way, to advance along the vertical fillet weld joining the web sheet with the girder end transverse stiffener, so that in the end the whole upper outer web corner tore away from the web peripheral frame (i.e., from the upper flange and the end stiffener). It was observed that the tearing off of the web sheet occurred in that portion of the web in which the diagonal tension band developed and was anchored into the boundary frame; see Fig. 2.9.

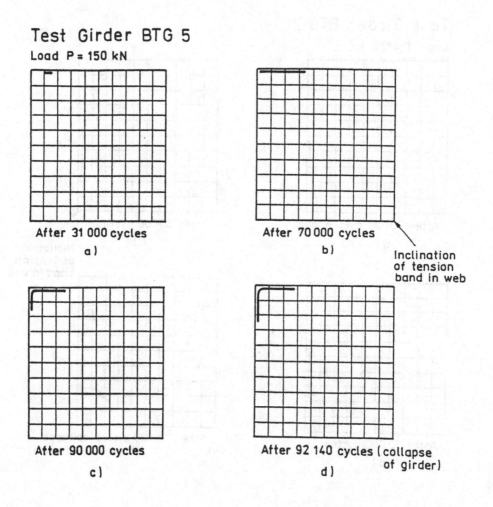

Fig. 2.9: Advance of a crack in the web of test girder BTG 5

(iv) OR in some cases, phenomenon (iii) interacted with phenomenon (ii), i.e. with a crack progressing (with a certain delay with respect to the first crack) along the transverse stiffener (Fig. 2.10).

The final stage of the failure mechanism of the test girder was then similar to that which can be observed in constant loading experiments on shear panels with openings in their webs. It again consisted of plastic hinges in the flanges and a plastic diagonal band in the web sheet, but their location and character were significantly influenced by the presence of a crack (i.e., of an "opening") in the sheet.

With one experimental girder (and it was one of the girders with thick flanges), no crack was detected; and the girder failed - under a load almost as high as the corresponding constant loading ultimate shear force, and very shortly after the related loading step was reached - in the usual way known from constant loading tests.

Test Girder BTG 6

Load P = 120 kN

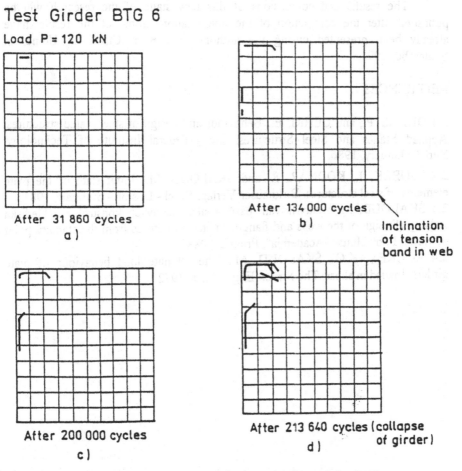

After 31 860 cycles
a)

After 134 000 cycles
b)

Inclination
of tension
band in web

After 200 000 cycles
c)

After 213 640 cycles (collapse
of girder)
d)

Fig. 2.10: Advance of a crack in the web of test girder BTG 6

2.2.3. Continuation of research

Further tests on the "breathing" of webs subject to repeated shear, comprising experiments on several new series of test girders, are being conducted at the author's Department in Prague. In each of these experiments, the girder is subject to a large number of the same loading cycles, which means that in the light of the classification given in the introduction to this section, the investigation is focused on the first group (i) of "breathing" problems.

The objectives of this research are twofold:

a) To establish S/N curves for webs under the action of repeated shear.

b) To experimentally determine stress ranges at those portions of the web in which fatigue cracks initiate and propagate, and therefore to obtain a basis for a subsequent theoretical study of the problem via the apparatus of Fracture Mechanics.

The results and conclusions of this new stage of the research will be published after the completion of the investigation. Some of the results have already been presented during the author's lectures at CISM, in Udine, in September 1994.

REFERENCES

2.1 DUBAS, P., and GEHRI, E., Behaviour and design of steel plated structures, Applied Statics and Steel Structures, Swiss Federal Institute of Technology, Zürich, January, 1986.

2.2 DJUBEK, J., KODNÁR, R. and ŠKALOUD, M.: Limit state of the plate elements of steel structures. Birkhäuser Verlag, Basel - Boston - Stuttgart, 1983.

2.3 ŠKALOUD, M. : Navrhování pásů a stěn ocelových konstrukcí z hlediska stability (Design of the webs and flanges of steel structures from the stability point of view). Publ. House "Academia", Prague, 1988.

2.4 ROCKEY, K.C., ŠKALOUD, M.: The ultimate load behaviour of plate girders loaded in shear. The Struct. Eng., No. 1, 1972

BEHAVIOUR OF PARTS OF STEEL FRAMES

M. Iványi

Technical University of Budapest, Budapest, Hungary

ABSTRACT

According to practical needs, approximate methods in stability analysis of steel structures are widely used, as information given by bifurcation theory is limited.

There are two main theories of plastic buckling, the first is based on Hencky's deformation theory, and the second, on the Prandtl–Reuss flow theory. First two chapters deal with these theories.

The complex of strain hardening of structural steel, residual stresses caused by technology, interaction of plate and lateral-torsional buckling significantly affects the stability conditions of beam columns. Theoretical results, obtained by the energy method relying on the theory of plastic deformation, are compared with test results in Chapter 3.

In Chapter 4, experimental and theoretical investigations are shown, which permit the analysis of plastic deformation capacity of frames with regard to the interaction of strength and stability phenomena, including the effect of plate (local) buckling on the response of the whole structure.

CHAPTER ONE

INTRODUCTION: BACKGROUND AND OBJECT

1.1 CRITICAL AND POST-CRITICAL BEHAVIOUR

A new aspect is coming into being in the methods of stability analysis of steel and metal constructions [1] [2]. The classical stability analysis based its methods of calculation on the occurence of bifurcation of equilibrium and the "critical" load parameter connected with the latter or on determination of stresses (P_{cR}, σ_{cr}). These quantities can be computed by exact or approximating formulas in closed form in easier instances and by the help of properly elaborated algorithms.

The content of information of critical load parameter is strongly limited. Essentially it only refers to the fact that unavoidable deviations (for example geometric inaccuracies) - which can be designated as "initial imperfections" - between real structure and model taken for calculation's basis, under the effect of critical load parameter, influence the result of calculation only to a limited degree; on the other hand, this statement is not valid in the vicinity of critical load parameter [3].

Because of this fact, critical load parameter could be used - considering other examinations of strength - only by the application of a rather large safety factor: this solution conduced to uneconomical results in part of the cases, and conduced to a still not adequately safe result in the other part of the cases [4].

The knowledge of critical load parameter in easier cases - first of all in case of bars - afforded a possibility of better estimation of deviations (mostly geometric-like) mentioned above with the introduction of $(1 - P/P_{cR})^{-1}$ formed magnification factor. The scope of application of this is also limited; besides it links to elastic region, besides it is connected with the aspect arising from the linearity of the problem that says that a construction can do unlimitedly large displacements by reaching of critical load parameter.

The non-linear examination of occurence of bifurcation [5] [6], which takes the occurence to stable, unstable and asymmetrical cases (Fig. 1.1), it enables the effect of deviations from ideal model of calculation to be better estimated - unfortunately only

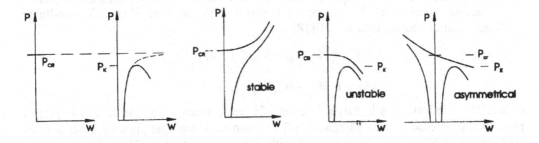

Figure 1.1

within the elastic region. The followings signify peculiar complications in case of local buckling of a plate.

This kind of bifurcation belongs to - in most cases - the kind, stable group: Because of it, there is an increasing post-critical way. Force–displacement diagrams of plates having initial disturbance (crookedness) are of the same type. Ultimate state of loading capacity at $\sigma_k = P_k / A$ average stress can be attributed in this way only to appearence of plastic regions and eventually to plastic instability. However, relation of σ_{cr} and σ_k is varying. In case of a thick plate (similarly to case of a rod): $\sigma_k < \sigma_{cr}$; reversely in case of a thin plate $\sigma_k > \sigma_{cr}$; moreover, $\sigma_k \gg \sigma_{cr}$ is possible (Fig. 1.2). In this way the magnification factor is invalid, furthermore σ_{cr} critical stress as base of comparison also loses its significance a lot [7]; real loading capacity only can be characterized by the analysis of post critical behaviour.

In this manner the calculation of loading capacity of a plate loaded by "initial"

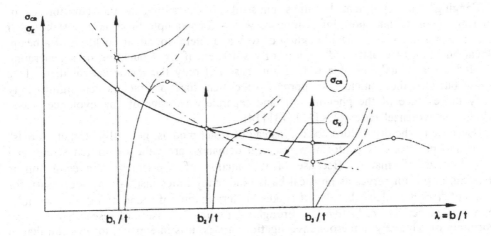

Figure 1.2

disturbances and doing finite displacements requires a non-linear, elastic-plastic analysis.

Starting formulas of this scope of problem - for instance Kármán's non-linear buckling equations established in 1910 [8] :

$$\Delta\Delta w = \frac{v}{D}\left(\phi_{yy}W_{xx} + \phi_{xx}W_{yy} - 2\phi_{xy}W_{xy}\right)$$

$$\Delta\Delta\phi = E\left(W_{xy}^2 - W_{xx}W_{yy}\right)$$

(where W is deflection, ϕ is Airy's function of stress), which were extended to imperfect plate in loaded condition by Marguerre [9] and also to orthotropic plate by other authors [10], and which also can be construed within the plastic region [11] - is known and certainly they are solvable with adequate computer technical instruments. Nevertheless, design methods based on "exact" following in the wake of ultimate state of loading capacity, apart from special instances, do not seem to be practical because of the followings:

(i) Scattering of certain fundamental parameters (crookedness, residual stresses, etc.) is relatively high and their statistical characterizations are mainly based on estimations: thus their inaccuracies do not fit in with the method's mathematical accuracy. Moreover, magnitude of initial geometric disturbances, and besides the variety of their possible shapes and diverse effect of their shapes cause more complication.

(ii) An examined plate (plate strip, plate range) is usually one building element (generating plate, flange, web-section) of some complete construction. Thus in so far as the behaviour of a building element can be only defined by large mathematical apparatus and usually only through numerical way, analysis of a complete construction becomes difficult. Hence the complicated analysis of local instabilitical occurence is hardly insertable in the compasses of global examination of the entire structure. .

(iii) The real behaviour of structural elements supporting a plate considerably and dominantly influences local buckling of a plate and mainly the post critical condition: the phenomenon only can be analyzed through the simultaneous examination of the plate and its "flanging". The problem becomes unusually embarrassing if the equilibrium of bordering elements can show bifurcation as well - for example there is a possibility of simultaneous occurence of local buckling of web and lateral-torsional buckling of a beam, of local buckling of a plate and bukling of a stiffener, of local buckling of a generating plate and column buckling of a rod. In this case not only the mathematical difficulties increase but also the coincide or approach of critical load parameters can significantly modify the essence of the phenomenon and separately pairings of "benevolence" cases produce "malevolence" ensembles [12] [13] [14].

Because of the difficulties above, instead of exact models, generally "target models" are put into practice that conform to practical demands, are valid in limited scope, and supply limited information. Because of it, function of experimental investigation is increasing, besides it serves as physical basis (and many times inspiration) neccessary for creation of "target models", besides, it makes the termination of round of validity possible. The later is especially productive if - through international division of research work - an opportunity of relatively numerous investigations arises. It is interesting to mention that in the seventies, the validity of eight target models elaborated for the examination of one of the acute problems of steel bridge building, the local buckling of an orthotropic floor plate was controlled by 105 rather large-scale (in this way extremely expensive) tests. During

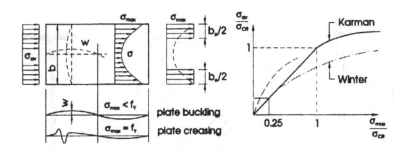

Figure 1.3

these experiments, considerations of mathematical statistics and probability theory were also applicable [15].

The applied target models essentially can be divided into two groups. The <u>first group</u> is suitable for examination of plates that are rather this and in this way having essential post critical reverse; and common feature of them is that they only take the membrane stresses for their basis during the calculation process of loading capacity, assuming those act henceforward in the original medium plane of the crooked plate. The starting thought originates from Kármán [16] in case of compression plates and from Wagner [17] in case of shear plates. The rather productive idea of the former was the introduction of the notion of post critical "effective plate width" (Fig. 1.3). With use of

$$\frac{b_e}{b} = \sqrt{\frac{\sigma_{cr}}{\sigma_{max}}} \quad \text{and} \quad \frac{b_e}{b} = \sqrt{\frac{\sigma_{cr}}{\sigma_{max}}}\left(1 - 0.25\sqrt{\frac{\sigma_{cr}}{\sigma_{max}}}\right)$$

- formed simple formulas which were created by the original one and by Winter on experimental basis (and thus reflecting the effect of "initial disturbances"), it affords a possibility of simple description of membrane stress rearrangement, the occurence of yielding at plate edge and of "crease" (post critical loss of loading capacity) resulting from this kind of yielding. Interpretation of experimental examination of compressed structures built up from quite thin plates is based upon this idea; common examination of column buckling and local buckling of compression rods of box cross sections [18] [19] [20] [21] [22] [23], simultaneous analysis of local buckling of web and lateral-torsional buckling [24].

Wagner's idea of post critical examination of a shear plate - accordingly in post critical state, a "tension strip" founded on inclined membrane stresses carries shear - has been extensively put into practice at structures on the basis of Basler's [25] suggestions and after a widespread experimental preparation it has developed into the new discipline of stiffened web plates' examination of local buckling.

The summary of Dowling and Chatterjee [15] has to be indicated in point of remarkably extensive examinations, emphasizing certain more important publications [26] [27] [28] [29] [30] [31] [32] [33]. One of the accepted target models of post critical loss of loading capacity calculation is shown in Fig. 1.4.

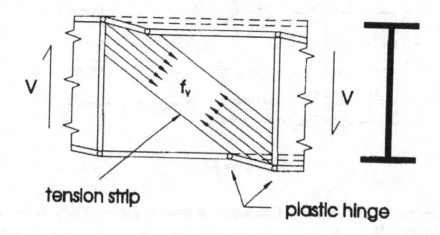

Figure 1.4

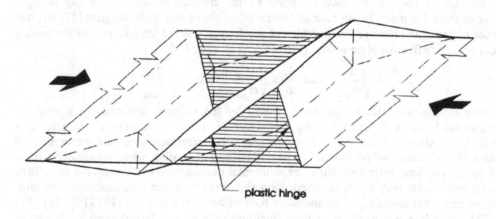

Figure 1.5

The <u>second group</u> of target models aimed the description of behaviour of relatively thick plates. Since in these cases the reverse of post critical loading capacity is relatively small (possibly there is not at all), practically local buckling comes into being with yielding at the same time. The fundamental question is the characterization of strains after local buckling in order that the simplified strain regularities of a structural element with "buckling", increasing or decreasing loading capacity could be used for the analysis of the entire structure. The same is suitable for fixation of criteria for election of thickness of a plate that makes the examination of occurence of local buckling omissible.

A feature of these target models is that they take also the function of bending moments occuring in a crooked plate into consideration besides membrane stresses. Simplification of examination can be done by the help of rigid-plastic "yielding mechanisms" - also applied at the plastic examination of loading capacity of rod assemblies -, (Fig. 1.5) [34]. On the basis of these mechanisms, at least the region of decreasing loading capacity, the strain condition of a plate can be characterized by a simple relation. These models can be extended to cases of elastic–strain hardening relations besides ideally elastic-plastic laws of material.

1.2 INTERACTION BETWEEN LOCAL AND OVERALL BUCKLING

The discussion so far has been concerned only with determination of the critical and postcritical loads. However, attainment of these loads are not in all cases synonymous with the occurence of collapse. Whether or not a section is capable of resisting further load above the critical one depends on the buckling mode concerned. If the lowest critical load corresponds to the overall mode, this constitutes an upper bound of the load bearing capacity. If the critical load corresponds to one of the other modes, determination of the maximum load is considerably more complicated. In such a case, the cause of collapse is usually an interaction between the different modes. It is a characteristic of this interaction that the constituent modes do not normally have the same wavelength. The process in the case of interaction can be relatively complex. A possible process will be discussed.

A column according to Fig. 1.6 is studied. It is assumed to be acted upon by an axial load centroid of the column. Application of this load is assumed to be such that the position of the load resultant is fixed during the loading sequence. It is further assumed, for the sake of simplicity, that the column is free of imperfections.

Initially, only a compression of the column will occur. For this case, it is assumed that the local buckling load constitutes the lowest critical load (bifurcation load). When this is exceeded, buckling of the flange occurs, with redistribution of stress as a result. (See Fig.1.6) At the same time, slight deflection of the web commences. This is primarily due to the fixing moment along the common edge between the wide flange and the web. However, uneven distribution of membrane stresses which occurs in consequence is of a lesser extent than in the wide flange. When load increases, greater redistribution of these gradually occurs in the web also. In simple terms, this occurs when the load on the web approaches the buckling load for a plate which has the same dimensions as the web.

In order to appreciate more easily what happens when local buckling occurs, Fig.1.7a,b may be studied. This shows the general relationship between load shortening in a plate which is acted upon by a compressive force **P**. The full curve refers to the ideal case (initially flat, perfectly elastic plate), while the dashed curve describes the sequence of events in a plate with imperfections.

It is seen in the figure that, in the ideal case, there is an abrupt reduction in tangent stiffness characterised by $d\sigma_m / d\varepsilon$ when the buckling load is reached. See Fig.1.7b. If, however, the plate has an initial deflection, there will be continual decrease in stiffness. This reduction in stiffnes occurs at a somewhat faster rate in the region of the critical buckling load. This is also the case when the load approaches the ultimate load.

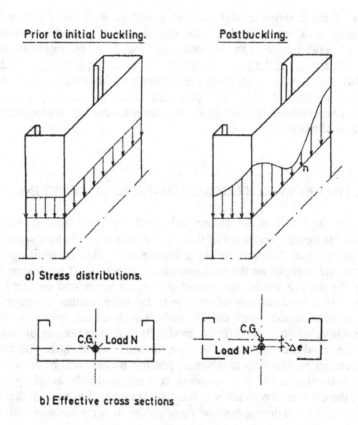

Prior to initial buckling.   Postbuckling.

a) Stress distributions.

C.G. Load N   C.G. Load N Δe

b) Effective cross sections

Figure 1.6: Stress distributions and effective cross sections before and after initial local buckling

There are now two possibilities, depending on the relation between the local buckling load and the Euler load for the column. If the local buckling load is near the Euler load, the rapid change in tangent stiffness in the event of local buckling may cause the residual bending stiffness of the column to become so low that overall buckling immediately occurs. If, on the other hand, the local buckling load is appreciably lower than the Euler load, deflection of the column is initiated when local buckling occurs. Further load can in this case be applied to the column before its load bearing capacity is exhausted. This occurs when, owing to buckling, the stiffness of the cross section has been reduced to such an extent that a stable equilibrium between external and internal forces is no longer possible. Reduction in stiffness can then occur either fully elastically as a result of local buckling, or due to a combination of this and partial plastification, which causes acceleration of the final collapse.

It will be evident from the foregoing discussion that the system is flexible while loading proceeds, and accurate analysis of the mode of action is usually extremely

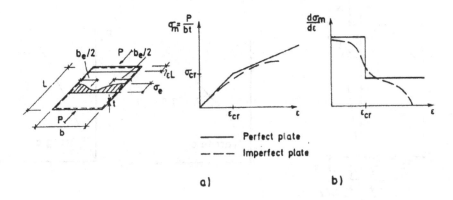

Figure 1.7: (a) Load–shortening curve for a compressed plate element (b) Corresponding tangent stiffness

complicated even when ideal conditions are assumed. If the effect of imperfections and plastic flow in the material must be taken into consideration, which is usually the case, the problem becomes even more complicated.

A more practical treatment of the problem is possible by applying the method which has been in use since the beginning of the 30's for dealing with plates in compression in the postcritical region. This is based on the effective width principle which is defined by the relation (see Fig. 1.7)

$$P = \int_0^b \sigma(y) \cdot t \cdot dy = \sigma_{av} \cdot b \cdot t = \sigma_{max} \cdot b_e \cdot t \qquad (1.1)$$

The effective width b_e is determined by a mathematical expression which is usually written as a function of the critical buckling stress σ_{cr} for the plate and the edge stress σ_{max}. By virtue of the fact that σ_{max} can be regarded as being proportional to the relative compression of the panel, equation (1.1) makes possible the determination of the relationship between load and compressive strain. This relationship is of fundamental importance in studying the interaction between local and overall buckling. Fig. 1.6b indicates how this problem can be tackled.

Croll and Walker [35] demonstrate how the spring model can be used to illustrate complex phenomena. Such a model can be used for a study in principle of the interaction between local and overall buckling. The model shown in Fig. 1.8a, where the behaviour of an axially loaded column whose flanges are represented by two springs are illustrated. The effect of the web is not considered separately. Formally, it can be regarded to be split into two and to form part of the two springs. The two springs are initially assumed to have the same spring constant c. It is further assumed, for the sake of simplicity, that the flange which is represented by spring No. 2 is stiffened in such a way that no local buckling is

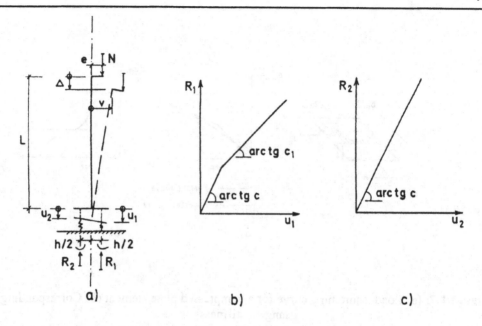

Figure 1.8: Spring model for study of the interaction between local and overall buckling.
(a) System sketch, notation (b) and (c) Spring characteristics

possible. This is reflected in the model by the fact that the spring constant c is all times applicable to spring No. 2. See Fig. 1.8c. In contrast, the flange represented by spring No.1 is assumed capable of local buckling. The load versus compressive strain curve for a plate in compression is shown in Fig.1.7 a. For the ideal case (perfect plate), this curve is in the form of two straight lines. Strictly speaking, the part of the curve which relates to the postcritical region is not straight but somewhat curved. Over a region immediately above the critical buckling load, however, the curve can be approximated by a straight line. A bilinear spring characteristic according to Fig. 1.8b is therefore assigned to spring No. 1.

In order to take into consideration the effect of initial curvature and plastic flow, the bilinear relation is replaced by a curve of the same general appearance as that in Fig.1.7a (imperfect plate). This system can be analysed by means of the equilibrium method. Deflections are assumed to be small. We then have (notations according to Fig. 1.8a)

$$N - R_1 - R_2 = 0 \tag{1.2}$$

$$N \cdot (e + v) + R_2 \frac{h}{2} - R_1 \frac{h}{2} = 0 \tag{1.3}$$

$$\frac{u_1 - u_2}{h} = \frac{v}{L} \tag{1.4}$$

$$R_1 = f(u_1) \tag{1.5}$$

$$R_2 = c \cdot u_2 \tag{1.6}$$

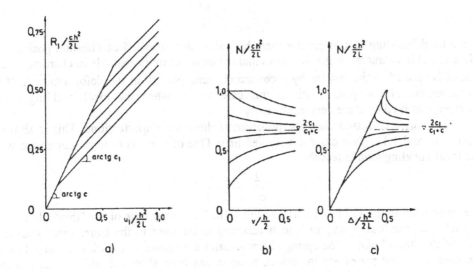

Figure 1.9: (a) Characteristics of spring 1 (b) Load–deflection curve N - v
(c) Load–displacement curve N - Δ

$$\Delta = \frac{u_1 + u_2}{2} + \frac{L \cdot (u_1 - u_2)^2}{2h^2} \tag{1.7}$$

It is best to solve this set of equations by selecting values of u_1 and determining N and v on the basis of the other conditions. It may be mentioned that two roots will be obtained for each selected value of u_1, of which it is only the smaller one which is of practical interest.

We find that

$$N = \frac{c \, h^2}{2 \, L} \tag{1.8}$$

constitutes an upper bound to the loadbearing capacity, and that this load is analogous to the Euler load for the real case.

The load–deflection curves for the cases described have been given in a non-dimensional form, using

$$\frac{N}{\dfrac{c \, h^2}{2 \, L}} \quad (\text{see} \ \frac{N}{N_E})$$

as the ordinate.

Fig. 1.9 shows the results, when conditions are assumed to be ideal, for different relationships between the local buckling load and Euler load. It is a characteristic of this system that deflection commences when either the local buckling load or the Euler load is reached. It is also found that the deflected equilibrium positions are of a stable character if the local buckling load is less than

$$\frac{2c_1}{c_1+c} \cdot \frac{c}{2} \cdot \frac{h^2}{L}$$

If the local buckling load is greater than this value, the deflected equilibrium positions are of an unstable character. Fig.1.9c shows that the equilibrium is unstable in character even if a load is applied to the system by a constraint being placed on the deformation Δ. It is a characteristic of collapse, which in this case occurs when the local buckling load is reached, that it takes place very explosively.

The mode of action changes, however, if there are imperfections. This is shown in Figs. 1.10 and 1.11. Two cases are studied here. One of these is to represent a case where the local buckling load is less than

$$\frac{2c_1}{c_1+c} \cdot \frac{c}{2} \cdot \frac{h^2}{L}$$

(i.e. is situated in the stable region), see Fig.1.10. This is the case of the "thin" plate. In the second case, the local buckling load is assumed to be more to the Euler load. This is the case of the "thick" plate. The spring characteristics are shown in Figs.1.10a and 1.11a. For the case when the plates are imperfect, a curve has been sketched which must represent qualitatively the actual load–compressive strain relation for a plate. This curve has then been used as the basis for calculation of the load–deflection relation. In both cases, the relations for ideal conditions are also shown. Imperfections in the column as a whole (overall imperfections), which are usually taken to include the initial curvature of the column and unintentional load eccentricity, are in this case assumed to be combined into a load eccentricity e.

In contrast to the ideal case, where deflection is initiated simultaneously with attainment of the bifurcation load, deflection in the case with imperfections usually occurs as soon as load is applied. Loss of stability occurs when the maximum is reached on the load–deflection curve. See Figs. 1.10b and 1.11b).

The load–deflection process is also greatly dependent on the type of imperfection. The occurence of imperfections is usually associated with a reduction in the load bearing capacity. That this is not generally true is shown in Fig. 1.10b, where the case $e/h = -0.1$ gives a higher maximum load than the case $e/h = 0$. This is due to the fact that, as a result of the direction of the load eccentricity, the buckling-prone plates is in this case subjected to a lesser load and final collapse is therefore delayed.

It is not an unusual phenomenon in the case of cold formed sections that some plates are unintentionally bent into a cylindrical shape during forming. Such a type of imperfection causes an elevation of the critical buckling load. Nor does the stiffness commence gradually to decrease as soon as load is applied, a phenomenon which is characteristic, for instance, in the event of initial buckling which has a lesser or greater affinity to the mode of buckling.

The object of the above section was to illustrate qualitatively, on the basis of a simple mechanical model, some of the problems encountered when there is interaction between local and overall buckling. The model reproduces the principal mode of action of a column which is subject to interaction between local and overall buckling. It is however far too simple to permit production of quantitative results which can be applied to the actual construction. The model particularly indicates a postcritical region of unstable character in cases where the local buckling load exceeds a certain limiting value

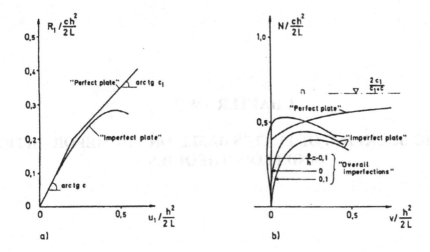

Figure 1.10: Examples of load–deflection curves when there are imperfections

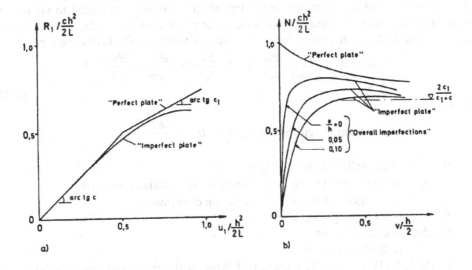

Figure 1.11: Examples of load–deflection curves when there are imperfections

$$\frac{2c_1}{c_1+c} \cdot \frac{c}{2} \cdot \frac{h^2}{L}$$

Owing to this property, the construction is particularly sensitive to imperfections, primarily when the local buckling load is of the same order as the Euler load [14]. This is the so-called "optimum erosion".

CHAPTER TWO

PLASTIC BUCKLING OF PLATES BASED ON THE DEFORMATION AND FLOW THEORIES

2.1 LITERATURE SURVEY

A thin plate may fail due to lateral deflection when the plate is subjected to compressive forces, shearing forces or their combination in its plane along the sides. The differential equation of such plates, originally derived by Saint-Venant, has the following form:

$$D\left(\frac{\partial^4 w}{\partial x^4} + 2\frac{\partial^4 w}{\partial x^2 \partial y^2} + \frac{\partial^4 w}{\partial y^4}\right) + t\left(\sigma_x\frac{\partial^2 w}{\partial x^2} + 2\tau_{xy}\frac{\partial^2 w}{\partial x \partial y} + \sigma_y\frac{\partial^2 w}{\partial y^2}\right) = 0$$

(2.1a)

where

D is the flexural rigidity of the plane: $D = \dfrac{Et^3}{12(1-v^2)}$

w is the deflection of the plate

σ_x, σ_y are the normal stress components in the cartesian coordinates

τ_{xy} is the shearing stress in the cartesian coordinates

E is the Young's modulus

t is the thickness of plate

v is the Poisson's ratio.

    In 1891, G. H. Bryan [36] investigated theoretically the phenomenon of buckling of rectangular plates which were simply supported on all edges and acted upon on two opposite sides by a uniformly distributed compressive force in the plane of the plate. He applied the energy criterion of stability to the buckling.

    More than fifteen years later, Timoshenko [37], Reissner [38], Sezawa [39], Wagner [40] and Taylor [41] treated problems concerning the buckling of rectangular plates fixed at edges and also under various boundary conditions. In particular, Timoshenko investigated extensively the stability of plates with various conditions of support under compressive forces, shearing forces or their combination.

An attempt to extend the theory of plate stability into the inelastic range was made by Bleich [42], by considering the plate which has a reduced modulus of elasticity in the direction of the loading, but which retains the Young's modulus in the direction perpendicular to the loading. The equilibrium equation in this case becomes:

$$D \left(\alpha_t \frac{\partial^4 w}{\partial x^4} + 2 \sqrt{\alpha_t} \frac{\partial^4 w}{\partial x^2 \partial y^2} + \frac{\partial^4 w}{\partial y^4} \right) + t \, \sigma_x \frac{\partial^2 w}{\partial x^2} = 0$$

(2.1b)

where $\alpha_t = E_t / E$, and E_t is the tangent modulus.

The middle term in the parantheses is associated with the distortion of a square element of the plate due to the twisting moment acting on the element. Taking into consideration the limiting values 1 and α_t, the coefficient of the middle term was assumed arbitrarily as $\sqrt{\alpha_t}$.

On the other hand, Chwalla [43], Ros and Eichenger [44] assumed that the plate is isotropic in the plastic range and hence that the modulus of elasticity is the same in both directions. The corresponding equilibrium equation becomes:

$$D \, \alpha_r \left(\frac{\partial^4 w}{\partial x^4} + 2 \frac{\partial^4 w}{\partial x^2 \partial y^2} + \frac{\partial^4 w}{\partial y^4} \right) + t \, \sigma_x \frac{\partial^2 w}{\partial x^2} = 0$$

where $\alpha = E_r / E$, and E_r is the reduced modulus.

From this equation the critical buckling stress is α_r times that in the elastic range.

As the theory of plasticity developed after 1940, new theories of plastic buckling were presented.

There were two main currents in this development, one based on Hencky's deformation theory [45] and the other on the Prandlt–Reuss flow theory [46] [47]. After 1940, Bijlaard [48] and Ilyushin [49] independently applied the deformation stress–strain relationship to the plate buckling problem and obtained a solution. The deformation theory was further developed by Stowell [50] who modified it based on Shanley's concept [52] which is explained below. The improved Stowell theory [51] showed a good agreement with experimental results. In contrast with the deformation theory, Handelmann and Prager [53] in 1948 presented the buckling theory of plates - in the plastic range, based on the flow theory. The flow theory, which is more complete than the deformation theory from the viewpoint of mathematical theory of plasticity, showed no resemblance to the experimental results. Pearson [54] improved the flow theory by using Shanley's concept. According to Shanley's concept of the tangent modulus theory, inelastic buckling of uniformly loaded members will occur at the tangent modulus load. It is assumed that there is no strain reversal in any part of the member at the instant of buckling, and that buckling of the member proceeds simultaneously with increase in load. This improved theory still gave much higher critical values when compared to experimental results.

Onat and Drucker [55] investigated the influence of initial imperfections on torsional buckling of a simplified model of a uniform cruciform cross section and showed that the flow theory leads to a reasonable correlation with experiment when unavoidable initial imperfections are taken into account. However, their conclusions have no application to the general study of plate buckling.

In 1955, Yamamoto presented the theory of plastic buckling of plates with consideration of the effect of an initial imperfection [56]. The theory predicts, approximately, the tangent modulus load and conelates reasonably with test results.

In the following year, Thurlimann and Haaijer developed the plastic buckling theory of plates, based on the flow theory [57]. Taking into account the initial imperfections, the four independent instantaneous flexure and shear moduli of an orthotropic plate were determined from the test results of the material under consideration. This theory gave good correlation with experimental results.

When plates are used as members of structures, such as built-up columns, shells of ship structures, plate girders and so on, the structures are quite often fabricated by welding. Residual stresses due to welding have presented problems concerning their influence on the strength of welded structures. Before the strength of welded structures could be studied, it was necessary to know the behaviour of materials due to welding.

Before 1936, the analytical and experimental investigation of the welded joint had presumed an elastic behaviour throughout the complete process of welding. Boulton and Lance Martin [58], as well as Rosenthal [59] simultaneously and independently showed, both analytically and experimentally, that welding induced plastic deformation of material in and near the weld, and that the residual stresses resulting after cooling were due to these plastic deformations. Further analytical and experimental investigations were made by Fujita [60] and Tall [61]. Experimental work was carried out by many other investigators: Griffits [62], Cordovi [63]. Okerblom [64] studied the deformation of welded metal structures and presented a theoretical method for the calculation of welding deformations.

In 1960 Okerblom presented a paper [65] concerning the influence of residual stresses on the stability of welded structures and structural members based on experimental results. His paper showed the possibility of elastic buckling of plate elements in the structure due to welding.

In the same year Yoshiki and others investigated analytically the influence of residual stresses on the elastic buckling of center welded plates with the aid of integral equations and showed that tha residual stresses could influence the elastic buckling strength of a plate, particularly in certain cases of residual stress distribution [66].

2.2 ELASTIC, ELASTIC-PLASTIC AND PLASTIC BUCKLING OF PLATES

2.21 Stress–Strain Relationship in the Elastic and the Plastic Ranges

When the deformation and stresses of plates are analyzed, the relationship between stress and strain must be defined both in the elastic and plastic ranges of the material. The most fundamental relation of stress and strain is that obtained from a coupon test in unaxial tension or compression. Fig. 2.1a shows a typical stress–strain relationship of a strain hardening material, and Fig. 2.1b presents an idealized stress–strain relationship (elastic perfectly plastic) for steel.

In the elastic range the well established theory of elasticity is based on the following relationship between stress and strain.

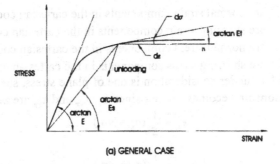

(a) GENERAL CASE

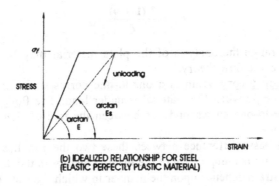

(b) IDEALIZED RELATIONSHIP FOR STEEL
(ELASTIC PERFECTLY PLASTIC MATERIAL)

Figure 2.1: Stress–strain relationship

$$\varepsilon_x = \frac{1}{E} \left(\sigma_x - \nu \left(\sigma_y + \sigma_z \right) \right)$$

$$\varepsilon_y = \frac{1}{E} \left(\sigma_y - \nu \left(\sigma_z + \sigma_x \right) \right)$$

$$\varepsilon_z = \frac{1}{E} \left(\sigma_z - \nu \left(\sigma_x + \sigma_y \right) \right)$$

$$\gamma_{yz} = \frac{2\left(1 + \nu\right)}{E} \, \tau_{yz}$$

$$\gamma_{zx} = \frac{2\left(1 + \nu\right)}{E} \, \tau_{zx}$$

$$\gamma_{xy} = \frac{2\left(1 + \nu\right)}{E} \, \tau_{xy}$$

(2.2a)

where

$\varepsilon_x, \varepsilon_y, \varepsilon_z$ are normal strain components in the cartesian coordinates

$\gamma_{yz}, \gamma_{zx}, \gamma_{xy}$ are shearing strain components in the cartesian coordinates

$\sigma_x, \sigma_y, \sigma_z$ are normal stress components in the cartesian coordinates

$\tau_{yz}, \tau_{zx}, \tau_{xy}$ are shearing stress components in the cartesian coordinates.

When the problem under consideration is one of plane stress, such as is the case with thin plates, this relationship becomes simpler since σ_z, τ_{yz} and τ_{zx} are zero:

$$\varepsilon_x = \frac{1}{E}(\sigma_x - v\sigma_y)$$

$$\varepsilon_y = \frac{1}{E}(\sigma_y - v\sigma_x)$$

$$\gamma_{xy} = \frac{2(1+v)}{E}\tau_{xy}$$

(2.2b)

In the plastic range the analysis of the plate material may be based on either the deformation theory or the flow theory.

The <u>deformation theory</u> assumes a one-to-one correspondence between stress and strain in the plastic range when the material is under load. The <u>flow theory</u>, on the other hand, assumes a one-to-one correspondence between the rate of change of stress and the rate of change of strain.

The important basic difference between these two theories lies in the fact that the stress–strain relationship is independent of the loading history in the deformation theory. In the flow theory, the strain depends upon the manner in which the state of stress is obtained. Although it appears logical that the loading history must play role, test results of Yoshiki, Heimerl and Pride have shown that only the deformation theory gives good results.

The secant modulus deformation theory assumes that the material is isotropic in the plastic range and that, the intensities of stress and strain are defined by the square root of the second invariant of the stress and strain tensors with constant factors that is for the state of plane stress.

$$\sigma_i = \sqrt{\sigma_x^2 + \sigma_y^2 - \sigma_x\sigma_y + 3\tau^2}$$

$$\varepsilon_i = \frac{2}{\sqrt{3}}\sqrt{\varepsilon_x^2 + \varepsilon_y^2 + \varepsilon_x\varepsilon_y + \frac{\gamma^2}{4}}$$

(2.3)

where σ_i is the intensity of stress, and ε_i is the intensity of strain.

Under proportional loading, the assumed isotropy of the material requires the relationship between stress and strain to be of such a form that

$$\frac{\sigma_x - v\sigma_y}{\varepsilon_x} = \frac{\sigma_y - v\sigma_x}{\varepsilon_y} = \frac{\tau_{xy}}{2(1+v)\gamma_{xy}} = \frac{\sigma_i}{\varepsilon_i} = E_s$$

(2.4)

where E_s is the secant modulus.

Accordingly, the stress–strain relationship during the process of loading is:

$$\varepsilon_x = \frac{1}{E_s} (\sigma_x - v\sigma_y)$$

$$\varepsilon_y = \frac{1}{E_s} (\sigma_y - v\sigma_x)$$

$$\gamma_{xy} = \frac{2(1 + v)}{E_s} \tau_{xy}$$

$$(2.5)$$

For unloading, the material is assumed to behave completely elastically and the relationship between stress and strain may be defined as follow:

$$d\varepsilon_x = \frac{1}{E} (d\sigma_x - v\,d\sigma_y)$$

$$d\varepsilon_y = \frac{1}{E} (d\sigma_y - v\,d\sigma_x)$$

$$d\gamma_{xy} = \frac{2(1 + v)}{E} d\tau_{xy}$$

$$(2.6)$$

where the relationship is given in the form of a variation to eliminate the effect of a permanent set.

Based on the flow theory, the stress–strain relationship for loading may be defined by the following equations

$$\dot{\varepsilon}_x = \frac{1}{E} (\lambda\dot{\sigma}_x - (v + \frac{\lambda - 1}{2})\,\dot{\sigma}_y)$$

$$\dot{\varepsilon}_y = \frac{1}{E} (-(v + \frac{\lambda - 1}{2})\dot{\sigma}_x + \frac{\lambda + 3}{4}\dot{\sigma}_y)$$

$$\dot{\varepsilon}_z = \frac{1}{E} (-(v + \frac{\lambda - 1}{2})\dot{\sigma}_x - (v - \frac{\lambda - 1}{4}\dot{\sigma}_y))$$

$$\dot{\gamma}_{xy} = \frac{2(1 + v)}{E} \dot{\tau}_{xy}$$

$$(2.7)$$

where

$\dot{\varepsilon}_x, \dot{\varepsilon}_y, \dot{\varepsilon}_z$ are the rate of change of strain components in the cartesian coordinates with respect to the independent parameter

$\dot{\sigma}_x, \dot{\sigma}_y, \dot{\sigma}_z$ are the rate of change stress components in the cartesian coordinates with respect to the independent parameter $\lambda = E / E_t$

The equations for unloading are:

$$\dot{\varepsilon}_x = \frac{1}{E} \left(\dot{\sigma}_x - v\dot{\sigma}_y \right)$$

$$\dot{\varepsilon}_y = \frac{1}{E} \left(\dot{\sigma}_y - v\dot{\sigma}_x \right)$$

$$\dot{\gamma}_{xy} = \frac{2 (1 + v)}{E} \dot{\tau}$$

$$(2.8)$$

2.2 Potential Energy in the Plate

The strain energy per unit volume which is stored during the deformation of the plate may be expressed by

$$dW(x,y,z) = \int_0^{\varepsilon_x} \sigma_x \, d\varepsilon_x + \int_0^{\varepsilon_y} \sigma_y \, d\varepsilon_y + \int_0^{\gamma_{xy}} \tau_{xy} \, d\gamma_{xy}$$

$$(2.9)$$

where W is the strain energy stored in the plate.

This is an equation valid both in the elastic range and in the plastic range. The actual evaluation of the strain energy will be performed by introducing the stress–strain relationships presented in Section 2.21 and by integrating dW over the entire volume of the plate.

The total potential energy is obtained in the form of a summation of the strain energy stored in the plate, W, and the work done by the external forces acting on the plate, U, with an arbitrary additive constant depending on the reference position. The value of this arbitrary constant may be selected by making the potential energy equation to zero for a suitable reference position. In stability problems, it is usually convenient to take the loaded state prior to buckling as such a reference position.

The potential energy at buckling, with the reference position as that prior to buckling, may be expressed for the following cases:

(a) in the elastic region, and (b) in the plastic region, where both the deformation and the flow theories are used.

(a) For the elastic part of the plate, the energy equation may be shown to be

$$V = \iint \frac{D}{2} \left(\left(\frac{\partial^2 w}{\partial x^2} \right)^2 + 2(1 - v) \left(\frac{\partial^2 w}{\partial x \partial y} \right)^2 + 2v \left(\frac{\partial^2 w}{\partial x^2} \right) \left(\frac{\partial^2 w}{\partial y^2} \right) + \left(\frac{\partial^2 w}{\partial y^2} \right)^2 \right) dx \, dy - $$

$$- \iint \frac{t}{2} \left(\sigma_x \left(\frac{\partial w}{\partial x} \right)^2 + 2\tau \left(\frac{\partial w}{\partial x} \right) \left(\frac{\partial w}{\partial y} \right) + \sigma_y \left(\frac{\partial w}{\partial y} \right)^2 \right) dx \, dy$$

$$(2.10)$$

where V is the potential energy of the plate.

(b) In contrast with the above one expression for the elastic part, the energy equations for the plastic part of the plate are presented below in two different forms. One is based on the secant modulus deformation theory and the other on the flow theory, which were described in Section 2.21.

Stowell improved Ilyushin's deformation theory by basing it on the Shanley concept that no strain reversal occurs in any part of the member under a uniform compressive load.

Equation (2.11) was derived in this study, and resembles Stowell's equation. The difference between Stowell's equation and Eq. (2.11) is in the coefficients c_3' and c_3''. In Stowell's equation $c_3' = c_3'' = 1 - \dfrac{3}{4} \dfrac{\sigma_x \sigma_y + 2\tau^2}{\sigma_i^2} \kappa$.

$$V = \iint \frac{D_d}{2} \left(c_1 \left(\frac{\partial^2 w}{\partial x^2}\right)^2 - c_2 \left(\frac{\partial^2 w}{\partial x^2}\right) \left(\frac{\partial^2 w}{\partial x \partial y}\right) + c_3' \left(\frac{\partial^2 w}{\partial x \partial y}\right)^2 + \right.$$
$$\left. + c_3'' \left(\frac{\partial^2 w}{\partial x^2}\right) \left(\frac{\partial^2 w}{\partial y^2}\right) - c_4 \left(\frac{\partial^2 w}{\partial y^2}\right) \left(\frac{\partial^2 w}{\partial x \partial y}\right) + c_5 \left(\frac{\partial^2 w}{\partial y^2}\right)^2 \right) dx\, dy - $$
$$- \iint \frac{t}{2} \left(\sigma_x \left(\frac{\partial w}{\partial x}\right)^2 + 2\, \tau \left(\frac{\partial w}{\partial x}\right) \left(\frac{\partial w}{\partial y}\right) + \sigma_y \left(\frac{\partial w}{\partial y}\right)^2 \right) dx\, dy$$

$$\tag{2.11}$$

where

$$c_1 = 1 - \frac{\sigma_x^2}{\sigma_i^2} (1 - v^2) \kappa$$

$$c_2 = \frac{4\, \sigma_x\, \tau_{xy}}{\sigma_i^2} (1 - v^2) \kappa$$

$$c_3' = 2 \left((1 - v) - \frac{\tau_{xy}^2}{\sigma_i^2} (1 - v^2) \kappa \right)$$

$$c_3'' = 2 \left(v - \frac{\sigma_x \sigma_y}{\sigma_i} (1 - v^2) \kappa \right)$$

$$c_4 = \frac{4 \sigma_y \tau_{xy}}{\sigma_i^2} (1 - v^2) \kappa$$

$$c_5 = 1 - \frac{\sigma_y^2}{\sigma_i^2} (1 - v^2) \kappa$$

$$\kappa = 1 - \frac{E_t}{E_s}$$

D_d is the flexural rigidity of plate in the plastic range based on the deformation

theory, $D_d = \dfrac{E_s t^3}{12(1 - v^2)}$

When the integral of Eq. (2.11) is applied to the elastic zone, the coefficients change and $E_t = E_{sec} = E$, so that the Eq. (2.11) coincides with Eq. (2.10).

Pearson introduced Shanley's concept into Handelman–Prager equation for plastic buckling of plates which were based on the flow theory. The following energy equation was derived in this study by using Pearson's considerations and the Handelman–Prager equation.

$$V = \iint \frac{D_f}{2}\left(c_1 \frac{\partial^2 w}{\partial x^2} + c_3' \frac{\partial^2 w}{\partial x \partial y} + c_3'' \frac{\partial^2 w}{\partial x^2} \frac{\partial^2 w}{\partial y^2} + c_5 \left(\frac{\partial^2 w}{\partial y^2}\right)^2 \right) dxdy -$$

$$\iint \frac{t}{2}\left(\sigma_x \left(\frac{\partial w}{\partial x}\right)^2 + 2\tau \frac{\partial w}{\partial x} \frac{\partial w}{\partial y} + \sigma_y \left(\frac{\partial w}{\partial y}\right)^2 \right) dxdy$$

(2.12)

where

$$c_1 = \frac{(1 - v^2)(3 + \lambda)}{(5 - 4v)\lambda - (1 - 2v)^2}$$

$$c_3' = 2(1 - v)$$

$$c_3'' = \frac{4(1 - v^2)(2v + \lambda - 1)}{(5 - 4v)\lambda - (1 - 2v)^2}$$

$$c_5 = \frac{4\lambda(1 - v^2)}{(5 - 4v)\lambda - (1 - 2v)^2}$$

D_f is the flexural rigidity of plate in the plastic range, based on the flow theory:

$$D_f = \frac{Et^3}{12(1 - v^2)}$$

Equation (2.12) may also be reduced to Eq. (2.10) for the elastic zone, since $E_t = E$.

The energy equations are easily handled in elastic-plastic problems by properly adjusting the coefficient of the strain energy for the elastic and plastic parts of the plate.

2.23 Theorem of Minimum Potential Energy and Equilibrium Differential Equation

From the theory of elasticity the theorem of minimum potential energy may be stated as (Sokolnikoff) [67]:

> "Of all displacements satisfying the given boundary conditions those which satisfy the equilibrium equation make the potential energy an absolute minimum."

Conversely, when the potential energy of the body is a minimum the body is in a state of equilibrium.

This is valid only for an elastic body, but not necessarily obeying Hooke's law. This fact implies that as long as the body is not subject to unloading in the plastic range this theorem is in effect. In this study the two theories used for solving the buckling problem are the secant modulus deformation theory and the flow theory; both are modified by Shanley's concept. Consequently it is assumed that there is no place in the plate which is subject to strain reversal at the instant of buckling and the theorem of minimum potential energy can be applied.

The fact that the theorem of minimum potential energy does lead to equilibrium differential equations in the plastic range (when no unloading occurs) was proved by comparing these equations with the equilibrium differential equations obtained from consideration of the equilibrium of an element of the body. This was done for both plasticity theories.

The differential equations of equilibrium of the plate in the plastic range are shown as follows for both plasticity theories:

(a) secant modulus deformation theory

$$c_1 \frac{\partial^4 w}{\partial x^4} - c_2 \frac{\partial^4 w}{\partial x^3 \partial y} + c_3 \frac{\partial^4 w}{\partial x^2 \partial y^2} - c_4 \frac{\partial^4 w}{\partial x \partial y^3} + c_5 \frac{\partial^4 w}{\partial y^4} =$$

$$= -\frac{t}{D_d} \left(\sigma_x \left(\frac{\partial^2 w}{\partial x^2} \right) + 2 \tau \left(\frac{\partial^2 w}{\partial x \partial y} \right) + \sigma_y \left(\frac{\partial^2 w}{\partial y^2} \right) \right)$$

(2.13)

where

$$\begin{aligned}
c_1 &= c_1 & \text{of Eq. (2.11)} \\
c_2 &= c_2 & " \\
c_3 &= c_3' + c_3'' & " \\
c_4 &= c_4 & " \\
c_5 &= c_5 & "
\end{aligned}$$

(b) flow theory

$$c_1 \frac{\partial^4 w}{\partial x^4} + 2 c_3 \frac{\partial^4 w}{\partial x^2 \partial y^2} + c_5 \frac{\partial^4 w}{\partial y^4} =$$

$$= -\frac{t}{D_f} \left(\sigma_x \left(\frac{\partial^2 w}{\partial x^2} \right) + 2 \tau \left(\frac{\partial^2 w}{\partial x \partial y} \right) + \sigma_y \left(\frac{\partial^2 w}{\partial y^2} \right) \right)$$

(2.14)

where

$$\begin{aligned}
c_1 &= c_1 & \text{of Eq. (2.12)} \\
c_3 &= c_3' + c_3'' & " \\
c_5 &= c_5 & "
\end{aligned}$$

The characteristic values of the above differential equations (2.13) and (2.14) give the exact values of buckling of strength.

In general, a calculation using this exact method for solution of the differential equations is much more involved than the energy method. If the solution of the differential equation is difficult to obtain, the energy method can be used as a powerful tool to solve th problem to sufficient accuracy for engineering purposes.

2.3 THEORETICAL DERIVATIONS

2.31 Definitions

The intensities of stress and strain are defined by Ilyushin and Stowell, respectively, as

$$\sigma_i = \sqrt{\sigma_x^2 + \sigma_y^2 - \sigma_x \sigma_y + 3\, \tau^2}$$

(2.15)

$$e_i = \frac{2}{\sqrt{3}} \sqrt{e_x^2 + e_y^2 + e_x e_y + \frac{\gamma^2}{4}}$$

(2.16)

where

σ_x is the stress in the x direction
ε_x is the strain in the x direction
σ_y is the stress in the y direction
ε_y is the strain in the y direction
τ is the shear stress
γ is the shear strain

According to the fundamental hypothesis of the theory of plasticity, the intensity of stress σ_i is a uniquely defined, single-valued function of the intensity of strain e_i for any given material if σ_i increases in magnitude (loading condition). If σ_i decreases (unloading condition), the relation between σ_i and e_i becomes linear as in purely elastic case.

In the equations of definition (2.15) and (2.16), the material is taken to be incompressible and Poisson's ratio = 1/2. The stress–strain relations compatible with the equations of definition (2.15) and (2.16) are:

$$e_x = \frac{\sigma_x - \dfrac{1}{2}\sigma_y}{E_{sec}} = \frac{S_x}{E_{sec}}$$

$$e_y = \frac{\sigma_y - \dfrac{1}{2}\sigma_x}{E_{sec}} = \frac{S_y}{E_{sec}}$$

$$\gamma = \frac{3\,\tau}{E_{sec}}$$

$$e_i = \frac{\sigma_i}{E_{sec}}$$

(2.17)

These relations imply isotropy of the material.

2.32 Variations of strain and stress

When buckling occurs, let ε_x, ε_y and γ vary slightly from their values before buckling. The variations $\delta\varepsilon_x$, $\delta\varepsilon_y$ and $\delta\gamma$ will arise partly from the variations of middle-surface strains and partly from starins due to bending; thus

$$\delta\varepsilon_x = e_1 - z\,\chi_1$$

$$\delta\varepsilon_y = e_2 - z\,\chi_2$$

$$\delta\gamma = 2\,e_3 - 2\,z\,\chi_3$$

(2.18)

in which e_1 and e_2 are middle-surface strain variations and e_3 is the middle-surface shear–strain variation, χ_1 and χ_2 are the changes in curvature and χ_3 is the change in twist, and z is the distance out from the middle surface of the plate.

The corresponding variations, δS_x, δS_y and $\delta\tau$ in S_x, S_y and τ must be computed. From equations (2.17),

$$S_x = E_{sec}\varepsilon_x$$

therefore

$$\delta S_x = E_{sec}\,\delta\varepsilon_x + \varepsilon_x\,\delta(\frac{\sigma_i}{e_i}) = E_{sec}\,\delta\varepsilon_x - \frac{\varepsilon_x}{e_i}\,(\frac{\sigma_i}{e_i} - \frac{d\sigma_i}{de_i})\,\delta e_i$$

(2.19)

Now the variation of the work of the internal forces is

$$\sigma_i\,\delta e_i = \sigma_i\,\delta\varepsilon_x + \sigma_y\,\delta\varepsilon_y + \tau\,\delta\gamma$$

so that

$$\delta e_i = \frac{\sigma_x\,\delta\varepsilon_x + \sigma_y\,\delta\varepsilon_y + \tau\,\delta\gamma}{\sigma_i} =$$

$$= \frac{\sigma_x\,e_1 + \sigma_y\,e_2 + 2\,\tau\,e_3 - z\,(\sigma_x\,\chi_1 + \sigma_y\,\chi_2 + 2\,\tau\,\chi_3}{\sigma_i}$$

(2.20)

Substitution of this value of δe_i in equation (2.19) gives

$$\delta S_x = E_{sec}\,\delta\varepsilon_x -$$

$$- \frac{\varepsilon_x}{\sigma_i\,e_i}\,(\frac{\sigma_i}{e_i} - \frac{d\sigma_i}{de_i})\,(\sigma_x\,e_1 + \sigma_y\,e_2 + 2\,\tau\,e_3 - z\,(\sigma_x\,\chi_1 + \sigma_y\,\chi_2 + 2\,\tau\,\chi_3)$$

Let the coordinate of the surface for which $\delta e_i = 0$ (the neutral surface) be $z = z_0$. The expression for z_0 is obtained by setting $\delta e_i = 0$ in equation (2.20):

$$z_0 = \frac{\sigma_x\,e_1 + \sigma_y\,e_2 + 2\,\tau\,e_3}{\sigma_x\,\chi_1 + \sigma_y\,\chi_2 + 2\,\tau\,\chi_3}$$

(2.21)

By introduction of this coordinate into the expression for δS_x and by recognition of σ_i / e_i as E_{sec} and $d\sigma_i / de_i$ as E_{tan}

$$\delta S_x = E_{sec} (\varepsilon_1 - z \chi_1) +$$

$$+ \frac{\varepsilon_x}{\sigma_i e_i} (E_{sec} - E_{tan}) (\sigma_x \chi_1 + \sigma_y \chi_2 + 2 \tau \chi_3) (z - z_0)$$

$$(2.22)$$

In similar way it may be shown that

$$\delta S_y = E_{sec} (\varepsilon_2 - z \chi_2) +$$

$$+ \frac{\varepsilon_y}{\sigma_i e_i} (E_{sec} - E_{tan}) (\sigma_x \chi_1 + \sigma_y \chi_2 + 2 \tau \chi_3) (z - z_0)$$

$$(2.23)$$

and

$$\delta \tau = \frac{2}{3} E_{sec} (\varepsilon_3 - z \chi_3) +$$

$$+ \frac{\gamma}{3 \sigma_i e_i} (E_{sec} - E_{tan}) (\sigma_x \chi_1 + \sigma_y \chi_2 + 2 \tau \chi_3) (z - z_0)$$

$$(2.24)$$

Variations of forces and moments. For the variations of the imposed forces T_x, T_y and T_{xy} and the moments M_x, M_y and M_{xy}

$$\delta T_x = \int_{-\frac{t}{2}}^{\frac{t}{2}} \delta \sigma_x \, dz$$

$$\delta T_y = \int_{-\frac{t}{2}}^{\frac{t}{2}} \delta \sigma_y \, dz$$

$$\delta T_{xy} = \int_{-\frac{t}{2}}^{\frac{t}{2}} \delta \tau \, dz$$

$$\delta M_x = \int_{-\frac{t}{2}}^{\frac{t}{2}} \delta \sigma_x z \, dz$$

$$\delta M_y = \int_{-\frac{t}{2}}^{\frac{t}{2}} \delta \sigma_y z \, dz$$

$$\delta M_{xy} = \int_{-\frac{t}{2}}^{\frac{t}{2}} \delta \tau z \, dz$$

$$(2.25)$$

where t is the thickness of the plate.

From equation (2.17), (2.22) and (2.23)

$$\delta M_x = \frac{4}{3} \int_{-\frac{t}{2}}^{\frac{t}{2}} (\delta S_x + \frac{1}{2} \delta S_y)\, z\, dz =$$

$$= \frac{4}{3} [E_{sec} (e_1 + \frac{1}{2} e_2) \int_{-\frac{t}{2}}^{\frac{t}{2}} z\, dz - E_{sec} (\chi_1 + \frac{1}{2} \chi_2) \int_{-\frac{t}{2}}^{\frac{t}{2}} z^2\, dz +$$

$$+ \frac{e_x + \frac{1}{2} e_y}{\sigma_i\, e_i} (E_{sec} - E_{tan})(\sigma_x \chi_1 + \sigma_y \chi_2 + 2\tau \chi_3) \int_{-\frac{t}{2}}^{\frac{t}{2}} (z^2 - z z_0)\, dz] =$$

$$= \frac{4}{3} \frac{E_{sec} t^3}{12} [-(\chi_1 + \frac{1}{2} \chi_2) + \frac{e_x + \frac{1}{2} e_y}{\sigma_i\, e_i} (1 - \frac{E_{tan}}{E_{sec}})(\sigma_x \chi_1 + \sigma_y \chi_2 + 2\tau \chi_3)$$

$$= - D' \{ [1 - \frac{3}{4} (\frac{\sigma_x}{\sigma_i})^2 (1 - \frac{E_{tan}}{E_{sec}})]\, \chi_1 +$$

$$+ \frac{1}{2} [1 - \frac{3}{2} \frac{\sigma_x \sigma_y}{\sigma_i^2} (1 - \frac{E_{tan}}{E_{sec}})]\, \chi_2 - \frac{3}{2} \frac{\sigma_x \tau}{\sigma_i^2} (1 - \frac{E_{tan}}{E_{sec}})\, \chi_3 \}$$

(2.26)

where

$$D' = \frac{E_{sec} t^3}{9}$$

Similarly,

$$\delta M_y = \frac{4}{3} \int_{-\frac{t}{2}}^{\frac{t}{2}} (\delta S_y + \frac{1}{2} \delta S_x)\, z\, dz =$$

$$= - D' \{ [1 - \frac{3}{4} (\frac{\sigma_y}{\sigma_i})^2 (1 - \frac{E_{tan}}{E_{sec}})]\, \chi_2 +$$

$$+ \frac{1}{2} [1 - \frac{3}{2} \frac{\sigma_x \sigma_y}{\sigma_i^2} (1 - \frac{E_{tan}}{E_{sec}})]\, \chi_2 - \frac{3}{2} \frac{\sigma_y \tau}{\sigma_i^2} (1 - \frac{E_{tan}}{E_{sec}})\, \chi_3 \}$$

(2.27)

$$\delta M_{xy} = \int_{-\frac{t}{2}}^{\frac{t}{2}} \delta \tau\, z\, dz =$$

$$= - \frac{D'}{2} \{ [1 - \frac{3\gamma^2}{\sigma_i^2} (1 - \frac{E_{tan}}{E_{sec}})]\, \chi_3 + \frac{3}{2} (\frac{\sigma_x \tau}{\sigma_i^2} \chi_2 + \frac{\sigma_y \tau}{\sigma_i^2} \chi_2)(1 - \frac{E_{tan}}{E_{sec}}) \}$$

(2.28)

In these expressions, the integrations of δS_x, δS_y and $\delta \tau$ in the plastic region have been taken over the entire thickness of the plate, with the implication that no part of the plate is being unloaded.

2.33 Equation of equilibrium

If $w(x,y)$ is the bending deflection of the plate at buckling, and if no external moments are applied to the plate, then the equation of equilibrium of an element of the plate may be written

$$\frac{\partial^2(\delta M_y)}{\partial x^2} + 2\frac{\partial^2(\delta M_{xy})}{\partial x \partial y} + \frac{\partial^2(\delta M_y)}{\partial y^2} = t\,(\sigma_x\,\frac{\partial^2 w}{\partial x^2} + \sigma_y\,\frac{\partial^2 w}{\partial y^2} + 2\tau\,\frac{\partial^2 w}{\partial x \partial y})$$

(2.29)

in which the imposed forces $\sigma_x t$, $\sigma_y t$ and τt are considered as given (σ_x and σ_y are positive for compression). In terms of w, the changes in curvatures are

$$\chi_1 = \frac{\partial^2 w}{\partial x^2}$$

(2.30a)

and

$$\chi_2 = \frac{\partial^2 w}{\partial y^2}$$

(2.30b)

The change in twist is

$$\chi_3 = \frac{\partial^2 w}{\partial x \partial y}$$

(2.30c)

When the values of δM_x, δM_y and δM_{xy} in equations (2.26), (2.27) and (2.28), respectively, are differentiated as required by equation (2.29) and substituted in that equation, the general differential equation of equilibrium for a plate in the plastic state is obtained as follows:

$$[1 - \frac{3}{4}\,(\frac{\sigma_x}{\sigma_i})^2\,(1 - \frac{E_{tan}}{E_{sec}})]\,\frac{\partial^4 w}{\partial x^4} - 3\,\frac{\sigma_x \tau}{\sigma_i^2}\,(1 - \frac{E_{tan}}{E_{sec}})\,\frac{\partial^4 w}{\partial x^3 \partial y} +$$

$$+ 2\,[1 - \frac{3}{4}\,\frac{\sigma_x \sigma_y + 2\tau^2}{\sigma_i^2}\,(1 - \frac{E_{tan}}{E_{sec}})]\,\frac{\partial^4 w}{\partial x^2 \partial y^2} - 3\,\frac{\sigma_y \tau}{\sigma_i^2}\,(1 - \frac{E_{tan}}{E_{sec}})\,\frac{\partial^4 w}{\partial x \partial y^3} +$$

$$+ [1 - \frac{3}{4}\,(\frac{\sigma_y}{\sigma_i})^2\,(1 - \frac{E_{tan}}{E_{sec}})]\,\frac{\partial^4 w}{\partial y^4} = -\frac{t}{D'}\,(\sigma_x\,\frac{\partial^2 w}{\partial x^2} + \sigma_y\,\frac{\partial^2 w}{\partial y^2} + 2\tau\,\frac{\partial^2 w}{\partial x \partial y})$$

(2.31)

In the elastic range, equation (2.31) reduces to the usual form

$$\nabla^4 w = -\frac{t}{D}\,(\sigma_x\,\frac{\partial^2 w}{\partial x^2} + \sigma_y\,\frac{\partial^2 w}{\partial y^2} + 2\tau\,\frac{\partial^2 w}{\partial x \partial y})$$

where

$$D = \frac{Et^3}{9}$$

2.34 Energy integrals

Equation (2.31) is the Euler equation that results from a minimization of the integral

$$\iint \left(\frac{D'}{2} \{ C_1 (\frac{\partial^2 w}{\partial x^2})^2 - C_2 \frac{\partial^2 w}{\partial x^2} \frac{\partial^2 w}{\partial x \partial y} + C_3 \left[(\frac{\partial^2 w}{\partial x \partial y})^2 + \frac{\partial^2 w}{\partial x^2} \frac{\partial^2 w}{\partial y^2} \right] - \right.$$

$$- C_1 \frac{\partial^2 w}{\partial x \partial y} \frac{\partial^2 w}{\partial y^2} + C_5 (\frac{\partial^2 w}{\partial y^2})^2 \} -$$

$$\left. - \frac{t \, \sigma_i}{2} \left[\frac{\sigma_x}{\sigma_i} (\frac{\partial w}{\partial x})^2 + \frac{2\tau}{\sigma_i} \frac{\partial w}{\partial x} \frac{\partial w}{\partial y} + \frac{\sigma_y}{\sigma_i} (\frac{\partial w}{\partial y})^2 \right] \right) dx \, dy$$

$$(2.32)$$

which represents the difference between the strain energy in the plate and the work done on the plate by the external forces. The coefficients in this integral are:

In the plastic region	In the elastic region
$C_1 = 1 - \frac{3}{4} (\frac{\sigma_x}{\sigma_i})^2 (1 - \frac{E_{tan}}{E_{sec}})$	$C_1 = 1$
$C_2 = 3 \frac{\sigma_x \tau}{\sigma_i^2} (1 - \frac{E_{tan}}{E_{sec}})$	$C_2 = 0$
$C_3 = 1 - \frac{3}{4} \frac{\sigma_x \sigma_y + 2\tau^2}{\sigma_i^2} (1 - \frac{E_{tan}}{E_{sec}})$	$C_3 = 1$
$C_4 = 3 \frac{\sigma_y \tau}{\sigma_i^2} (1 - \frac{E_{tan}}{E_{sec}})$	$C_4 = 0$
$C_5 = 1 - \frac{3}{4} (\frac{\sigma_y}{\sigma_i})^2 (1 - \frac{E_{tan}}{E_{sec}})$	$C_5 = 0$

If there is a restraint of magnitude e along one longitudinal edge of the plate, the strain energy in this restraint itself is taken to be

$$\frac{e}{2} \frac{D}{b} \int [(\frac{\partial w}{\partial y})_{y=y_0}]^2 \, dx$$

$$(2.33)$$

Here y_0 is the edge coordinate. In expression (2.33), the stiffness D' is assumed to be the same as that in equation (2.26). If restraints are present along two edges, there will be two terms similar to expression (2.33). These terms may be added to integral (2.32) as additional strain energy.

2.35 Critical stress in plastic region

If the integral (2.32), supplemented if necessary by additional terms of the form of expression (2.33), is set equal to zero and the resulting equation solved for σ_x, the critical stress intensity in the plastic region $(\sigma_i)_{nl}$ is

$$(\sigma_i)_{pl} = \frac{\{\iint C_1(\frac{\partial^2 w}{\partial x^2})^2 - C_2\frac{\partial^2 w}{\partial x^2}\frac{\partial^2 w}{\partial x \partial y} + C_3[(\frac{\partial^2 w}{\partial x \partial y})^2 + \frac{\partial^2 w}{\partial x^2}\frac{\partial^2 w}{\partial y^2}] - C_4\frac{\partial^2 w}{\partial x \partial y}\frac{\partial^2 w}{\partial y^2} + C_5(\frac{\partial^2 w}{\partial y^2})^2\} \, dx \, dy + \frac{e}{b}\int[(\frac{\partial w}{\partial y})_{y=y_0}]^2 \, dx}{\iint[\frac{\sigma_x}{\sigma_i}(\frac{\partial w}{\partial x})^2 + 2\frac{\tau}{\sigma_i}\frac{\partial w}{\partial x}\frac{\partial w}{\partial y} + \frac{\sigma_y}{\sigma_i}(\frac{\partial w}{\partial y})^2] \, dx \, dy}$$

(2.34)

in which the values of the C's in the plastic range are used. This expression for the critical stess intensity may be minimized as with the corresponding elastic case.

If the values of the C's in the elastic region are used in formula (2.34), the critical stress intensity in the elastic region $(\sigma_i)_{el}$ is as follows:

$$(\sigma_i)_{el} = \frac{D}{t}\frac{\iint[(\frac{\partial^2 w}{\partial x^2})^2 + (\frac{\partial^2 w}{\partial x \partial y})^2 + \frac{\partial^2 w}{\partial x^2}\frac{\partial^2 w}{\partial y^2} + (\frac{\partial^2 w}{\partial y^2})^2] \, dx \, dy + \frac{e}{b}\int[(\frac{\partial w}{\partial y})_{y=y_0}]^2 \, dx}{\iint[\frac{\sigma_x}{\sigma_i}(\frac{\partial w}{\partial x})^2 + 2\frac{\tau}{\sigma_i}\frac{\partial w}{\partial x}\frac{\partial w}{\partial y} + \frac{\sigma_y}{\sigma_i}(\frac{\partial w}{\partial y})^2] \, dx \, dy}$$

(2.35)

CHAPTER THREE

LOCAL BUCKLING AND FLEXURAL-TORSIONAL BUCKLING WITH REGARD TO RESIDUAL STRESSES

3.1 INTRODUCTION

The analysis of plastic load-carrying capacity of steel structures requires to meet a system of conditions of several subsystems referring either to material and geometry of the structure, to residual stresses arising in manufacture, or to ways of loading, etc. A significant part of the system of conditions refers to stability phenomena, and to interaction between stability and strength phenomena.

This study mainly concerns the interaction between plate and lateral-torsional buckling, as well as some factors of significance for the interaction mentioned.

3.2 EXPERIMENTAL INVESTIGATIONS

Experimental investigations are essentially expected partly to supply physical background (often inspiration) needed to establish models for theoretical examinations, and partly to delimit the range of validity.

The complex series of experimental investigations involved measurements to determine:

(a)   plastic material characteristics of structural steel;
(b)   residual deformations;
(c)   load carrying capacities of members affected by compression, by bending and by eccentric compression.

3.21 Plastic material characteristics of steel

These have been determined in tensile tests on a total of 76 standard specimens of different plate thicknesses. Material testing results of web and flange plates of specimen No. 21 are seen in Fig. 3.1.

3.22 Residual deformations

Residual strains or stresses of welded I sections have been determined by the sectioning method. Residual strains affect both strength and stability phenomena. Tests comprised determination of residual deformations of 32 different cross-sections in all. The measured distribution of residual stresses in specimen No. 21 is seen in Fig. 3.2.

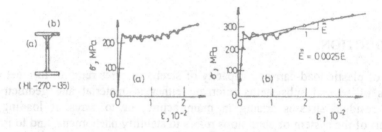

Figure 3.1: Material testing results of web and flange plates of specimen No. 21

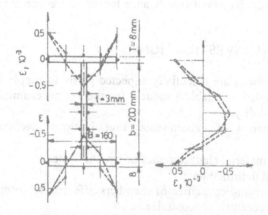

Figure 3.2: Measured distribution of residual strains in specimen No. 21

3.23 Load carrying capacities

Load carrying capacities of members in compression, bending and eccentric compression have been tested by experimental methods. In the case of eccentric compression, specimens were subjected to a constant compression force equal to 3 or 3 times the plastic compression capacity and the moment was increased step by step. The testing programme covered I sections of two different flange widths: web height ratios (**B / b = 0.6 and 0.8**). Each of the two cases had four different web slendernesses, each of them with four different flange widths (Fig. 3.3). Thereby a series comprised 32 specimens in all; each type was influenced by four different distributions of stresses. The experimental program involved testing of 132 specimens altogether.

3.3 PHENOMENA AND EFFECTS

Strain hardening of steel has been determined from the uniaxial state of stresses. The condition of uniaxial stress state for an elastic–ideally plastic material is illustrated in Fig. 3.4a, and of an elastic–strain hardening material in Figs. 3.4b-c.

Equations describing the material behaviour in the plastic range may be established based either on Hencky's theory of plastic deformations, or on the Prandtl–Reuss theory of plastic yield [68].

For strain hardening material, the theory of plastic deformations yields the material law of a non-linear elastic solid. The theory of plastic deformations cannot be generally considered as perfect, and the relevant results can only be accepted as approximations, but in the case of a so-called 'simple load', if, upon loading, stress components at any point of the solid grow in proportion to some parameter, the material equations of Hencky yield theoretically exact results [69].

Distribution and intensity of residual stresses depend on several factors, including geometry, material quality, various manufacturing effects, technological processes (rolling,

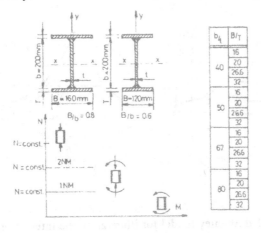

Figure 3.3: Variation of geometrical properties and loading conditions during the test program

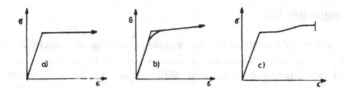

Figure 3.4: Simplified models for material: (a) elastic–ideally plastic; (b) and (c)
elastic–strain hardening

welding, straightening, etc.).

Residual stresses are mainly accessible to experimental methods [70]. Experimental
investigations involve technical difficulties; at the same time, quantitative and qualitative
(distributional) characteristics of residual stresses impose statistical analyses.

The problem of interactions is illustrated by analysing the interaction between plate
and lateral-torsional buckling on a simple example involving the behaviour of the so-called
Shanley model [71] (Fig. 3.5). The model is that of a rigid member subjected to a
compression force; the supporting springs simulate the behaviour of bar flanges, and the
effect of the web plate is neglected.

Spring No. 2 has a special (bilinear) characteristic curve taking the post-critical
behaviour of the flange plate into consideration after equilibrium bifurcation. Up to a
sufficiently high force level measuring P_L at flange buckling, both springs behave
identically and the so-called Euler (critical) load of the model can be expressed as

$$P_E = \frac{b^2}{2h}c \tag{3.1}$$

However, for $P_L < P_E$, at a force $P = P_L$, horizontal deflection (δ) will continuously
increase, and the force–displacement curve of the perfect model leads to the force
determined by spring factors c and c_i:

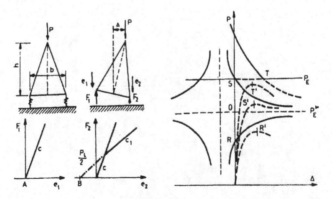

Figure 3.5: The so-called Shanley model for illustrating the interaction between plate and
lateral-torsional buckling

$$P_E^* = \frac{b^2}{2h} \frac{2cc_1}{c+c_1} \tag{3.2}$$

At force $P_L = P_E^*$, a critical state of equilibrium develops (Engesser–Kármán's critical load).

Taking various P_L forces in the relationship above into consideration, different load–displacement curves can be determined.

For $P_L < P_E^*$, equilibrium bifurcation of the perfect model is at point R. For an imperfect model, this condition involves a curve reaching to peak R'. This is the case of the "thin" plates. Imperfect characteristics are due to geometric imperfections, and to the occurrence of residual stresses.

For $P_E^* < P_L < P_E$, the post-critical state is of unstable character, with the possibility of significant differences between load carrying capacity values of perfect and imperfect models. For $P_E = P_L$, member and plate buckling are simultaneous. Points S and S' greatly differ, giving rise to the so-called "optimum erosion" [72].

Provided $P_E < P_L$, plate buckling affects only the condition of the perfect model after equilibrium bifurcation due to "lateral-torsional" bar buckling. Duly selecting the parameters, points T and T' may be made to belong to nearly identical load carrying capacities, and the imperfect model endowed with deformation capacity; this is the case of "thick" plates. Thereby, in plastic design, the influence of plate buckling on the deformation capacity of a structural part or of the complete structure becomes conspicuous.

3.4   INTERACTION BETWEEN PLATE AND LATERAL-TORSIONAL BUCKLING

Recently, interaction between plate and lateral-torsional buckling has come to the foreground of interest. Analyses have been made using different models and methods. One solution of the problem applied the theory of folded plates [73] [74]. The most current methods are the energy and the variational ones [75] [76]. In the early 1970s, the finite element method [77] [78], then the finite strip method [79] [80] have been introduced for the analysis of the interaction between plate and lateral-torsional buckling.

This chapter presents an analysis based on the energy method, taking also the plastic state of the steel material into consideration [81] [82]. The member is exposed to normal force and/or to constant bending moment; the effect of residual stresses in the bar due to the manufacturing technology is also taken into calculation.

The general geometry of the tested members is shown in Fig. 3.6. Specimens are simply supported at both ends with regard to both bending and twisting.

The member material is assumed to be strain hardening (Fig. 3.7); provided that residual stresses σ_r are lower than the yield stress σ_y, the relationship on Fig. 3.7a will be applied. Otherwise that in Fig. 3.7b is used. The transition between the elastic and the strain hardening ranges may be written in terms of a polynomial, assuming the residual strains to be of linear distributions.

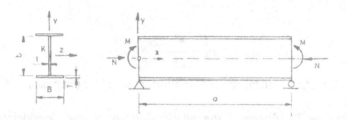

Figure 3.6: General geometry of tested members

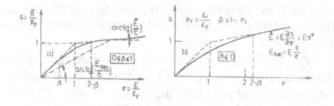

Figure 3.7: Elastic–strain hardening material model for calculations, if residual stress is lower (a) or higher (b) than yield strain

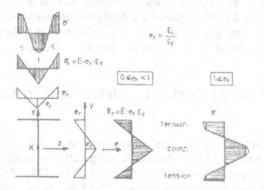

Figure 3.8: Supposed distribution of residual strains (stresses) due to technology processes over the cross-section

Distribution of residual strains and stresses due to technology processes over the cross-section is accounted for according to Fig. 3.8. Residual strains are assumed to be of linear distribution both in the flanges and in the web; residual stresses arising in the cross-section will be referred to this strain diagram of the elastic–strain hardening material.

Until the development of the critical state, strains due to loading will be considered to be of constant distribution in the flanges, and varying linearly in the web (Fig. 3.9a). Thereby the loading can be described in terms of the strain in the most compressed extreme fibre and parameter α, and residual strains added, making up the load seen in the cumulative diagram in Fig. 3.9b.

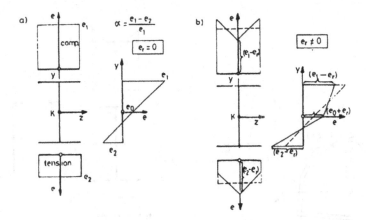

Figure 3.9: Strain distributions (a) from external loading, (b) resultant of (a) and residual strains

Taking the α value constant in an actual analysis, a so-called simple load is obtained: pure compression for $\alpha = 0$, pure bending for $\alpha = 2$, while $0 < \alpha < 2$ means a combined effect of compression and bending.

The analysis of the investigated phenomenon uses the theory of plastic deformations.

Taking into consideration the so-called Shanley phenomenon (i.e. at equilibrium bifurcation in the plastic range unloading parts, i.e. parts in which the previously achieved influences are decreasing, do not exist in the structure) into consideration, the potential energy of the member can be summed from those of the web and flange plates.

Relying on the results of Bijlaard and Ilyushin, Stowell [83] has established the potential energy of the web plate in the form

$$U_w = \frac{1}{2} \int_0^a \int_{-b/2}^{b/2} D' \cdot \left[C_1 (w'')^2 + (\dot{w}')^2 + w'' \ddot{w} + (\ddot{w})^2 \right] dx dy \qquad (3.3a)$$

$$V_w = -\frac{1}{2} \int_0^a \int_{-b/2}^{b/2} t \cdot \sigma_w \cdot (w')^2 \, dx dy \qquad (3.3b)$$

in which

$$\sigma_w = E \cdot \varepsilon_y \cdot s_w$$

$$C_1 = \frac{1}{4} + \frac{3E_t}{4E_{sec}} = \frac{s_w + 3s_w^* \cdot e_w}{4s_w}$$

$$D' = \frac{E_{sec} t^3}{12(1 - v^2)} = E \frac{s_w t^3}{9e_w} \qquad (v = 0.5)$$

$$w' = \frac{\partial w}{\partial x}; \qquad \dot{w} = \frac{\partial w}{\partial y}; \qquad s_w^* = \frac{\partial s_w}{\partial e_w}$$

The potential energy of the flanges can also be written as follows. If at $y = \pm b / 2$ in the extreme fibres $\varphi = -(w')$, then

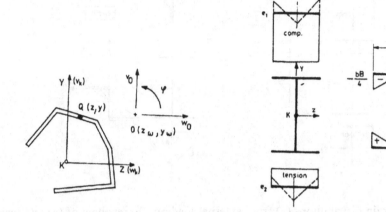

Figure 3.10: Relation between the displace-   **Figure 3.11:** Illustration for torsion
ments of centroid (K) and torsion centre (O)   centre migration in the plastic range

$$U_f = \frac{1}{2}\int_0^a \int_{A_f} E_{sec}\left[\omega_0^2(\varphi'')^2 + (w_0'')^2 z^2\right]dA_f dx + \frac{1}{2}\int_0^a (\varphi')^2 \int_{-B/2}^{B/2} G_{sec}\frac{T^3}{3}dzdx \quad (3.4a)$$

$$V_f = \frac{1}{2}\int_{A_f}(\sigma_{f1} - \sigma_{f2})\int_0^a\left[(w_k')^2 - 2yw_k'\varphi' + (y^2 + z^2)(\varphi')^2\right]dA_f dx \quad (3.4b)$$

in which

$$\sigma_{f1} = E\cdot\varepsilon_y\cdot s_1 \text{ and } \sigma_{f2} = E\cdot\varepsilon_y\cdot s_2$$

$$G_{sec} = \frac{E_{sec}}{2(1+v)} = \frac{E_{sec}}{3} \qquad (v = 0.5)$$

The overall potential energy of the member can be summarized as:

$$\Pi = U_w + W_w + U_f + W_f \qquad (3.5)$$

A cross-section of a general shape is seen in Fig. 3.10; its centroid is designated by K and the torsion centre by O. Displacements of point Q can be expressed as

$$w_Q = w_0 + \varphi\cdot(y_\omega - y) = w_K - \varphi\cdot y$$
$$v_Q = v_0 - \varphi\cdot(z_\omega - z) = v_K + \varphi\cdot z \qquad (3.6)$$

For a doubly-symmetric cross-section in the elastic range, centroid and torsion centre are coincident.

In the plastic range, these two points do not coincide. Even for a doubly-symmetric I section (as seen in Fig. 3.11), the following equilibrium equations should be fulfilled:

$$\int_{A_r} \sigma_\omega^{pl} dA_f = \varphi'' \int_{A_r} E_{sec} \omega_0 dA_f \quad = 0$$

$$\int_{A_r} \sigma_\omega^{pl} y dA_f = \varphi'' \int_{A_r} E_{sec} \omega_0 y dA_f \quad = 0 \tag{3.7}$$

$$\int_{A_r} \sigma_\omega^{pl} z dA_f = \varphi'' \int_{A_r} E_{sec} \omega_0 z dA_f \quad = 0$$

Because of the symmetry with respect to the vertical axis of the section, $\omega_0 = 0$ and $z_\omega = 0$, while

$$y_\omega = \frac{\int_{A_r} E_{sec} \omega_k z dA_f}{\int_{A_r} E_{sec} z^2 dA_f} \neq 0$$

and

$$\omega_0 = \omega_k - y_\omega z$$

The effect of torsion centre migration (change of y_ω) in the plastic range as a function of plastic deformation has been taken into account in writing potential energies of flanges.

The critical state can be determined by varying the function $w(x,y)$ so as to give a minimum for the function Π. The problem has been solved by the Ritz method.

Let the deformed shape be:

$$w(x,y) = X(x) \cdot Y(y) \tag{3.8}$$

Taking the boundary conditions of the member into consideration, the $X(x)$ function can be chosen

(a) for a member with pinned ends with regard to twisting:

$$X(x) = \sum_p \sin \frac{p \pi x}{a} \tag{3.9a}$$

(b) for a member with fixed ends with regard to twisting:

$$X(x) = \frac{1}{2} \sum_p \left(\cos \frac{(p-1)\pi x}{a} - \cos \frac{(p+1)\pi x}{a} \right) \tag{3.9b}$$

Possible deformations of the cross-section are shown in Fig. 3.12; thus they can be expressed as:

$$Y(y) = \sum_p A_{pq} \sin \frac{q\pi}{b} \left(y + \frac{b}{2} \right) + C_p \frac{y}{b} + D_p \tag{3.10}$$

Also, according to Fig. 3.12, the cases of plate buckling and of lateral-torsional buckling may be analysed separately by means of coefficients A_{pq}, C_p and D_p (see details in Section 3.6).

Substituting derivatives of function $w(x,y)$ into the potential energy term Π, the specific relative strain value e_1 leading to the critical state for a given α can be found. This problem has been solved by a PDP 11/34 computer. The program selects as many terms from the series of functions $w(x, y)$ as needed for a deviation below 2% between values for the nth and (n + 1)th terms in the final result. In the cases examined, nine terms always sufficed.

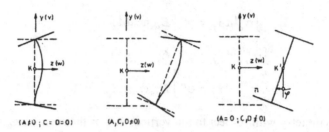

Figure 3.12: Deformations of the cross-section in cases of (a) web plate buckling; (b) combined web plate buckling and lateral-torsional buckling; (c) lateral-torsional buckling

3.5   COMPARISON OF THE RESULTS OF THEORETICAL AND EXPERIMENTAL INVESTIGATIONS

In the theoretical computations, geometrical imperfections and yield stress differences due to different plate thicknesses have been taken into consideration. Test results have been compared with load carrying capacities obtained on a rigid–ideally plastic material model (Fig. 3.13).

Residual strains of welded I sections have been determined, and measured values utilised in theoretical analyses.

In the plastic range, structural steel has a real yield plateau and strain hardening, which has been approximated by a single, substitutive strain hardening modulus in the analysis.

Results are seen to agree in Figs. 3.14 and 3.15, which summarize the results on specimens in pure compression (Series `N') and in pure bending (Series `M') for a flange width to web depth ratio, **B / b**, of **0.8**. A similar agreement was found between experimental and theoretical results on specimens subjected by eccentric compression (Series `1NM' and `2NM').

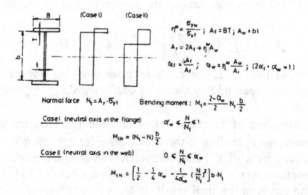

Figure 3.13: Theoretical load carrying capacities obtained on a rigid–ideally plastic material model

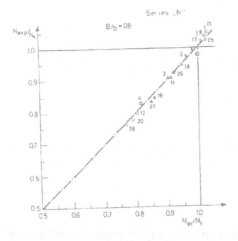

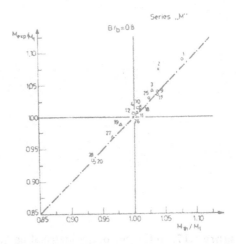

Figure 3.14: Comparison of experimental and theoretical load carrying capacities for specimens in pure compression

Figure 3.15: Comparison of experimental and theoretical load carrying capacities for specimens in pure bending

Load carrying capacity ratios depending on component plate slenderness (**B / T** and **b / t** being width to thickness ratios of flange and web, respectively) can be seen in Figs. 3.16 and 3.17, which indicate plate parameters, where the load carrying capacity determined from buckling coincides with that for the rigid–ideally plastic state. For an axial load, the flange slenderness prevails over the web plate slenderness (Fig. 3.16), while in pure bending (Fig. 3.17) both slenderness have similar influence.

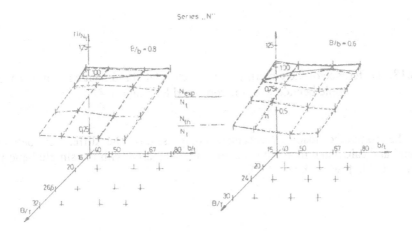

Figure 3.16: Influence of geometrical parameters on load carrying capacities of compressed members

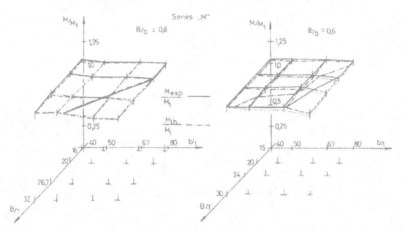

Figure 3.17: Influence of geometrical parameters on load carrying capacities of bended members

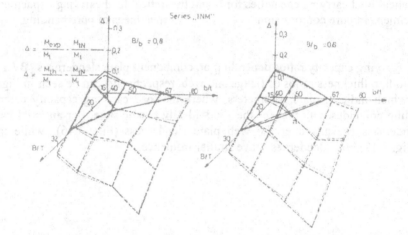

Figure 3.18: Difference of the experimental and calculated moment capacities compared to simple bended case for combined loading ($N = N_t / 3$)

For the combined loading (eccentric compression), the difference between the experimental and calculated moment capacities are compared with the simple bending case and are presented in Figs 3.18 and 3.19.

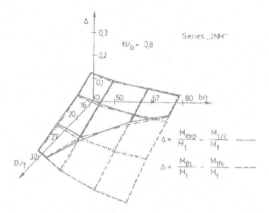

Figure 3.19: Difference of the experimental and calculated moment capacities compared to simple bended case for combined loading $(N = 2N_t / 3)$

3.6 PARAMETRIC STUDY OF INTERACTION

The theoretical method, experimentally qualified as adequate, has been applied to examine plate buckling, lateral-torsional buckling, and their interaction, for a member of typical cross-section (Fig. 3.20).

The test piece was supported as a hinged member for bending, and as a fixed member for twisting. Residual strains have been assumed to have a distribution as shown in the figure, with values compared with the yield strain ($e_r = \varepsilon_r / \varepsilon_y$).

The load has been described in terms of the relative strain of the extreme fibre in compression ($e_1 = \varepsilon_1 / \varepsilon_y$) and of parameter $\alpha = 1 - e_2 / e_1$, to be determined from the ratio of the extreme fibre strains ($\alpha = 0$ for pure compression, $\alpha = 2$ in pure bending).

Analyses referred to the effects of member length, the strain hardening modulus of the steel material, residual stresses and cross-sectional deformations allowing web buckling (when $A_{pq} \neq 0$ and $C_p = D_p = 0$), lateral-torsional buckling (when $A_{pq} = 0$ and $C_p \neq 0$ and $D_p \neq 0$) or the combined case (when $A_{pq} \neq 0$, $C_p \neq 0$ and $D_p \neq 0$).

Granting the simultaneous possibility of web buckling and lateral-torsional buckling,

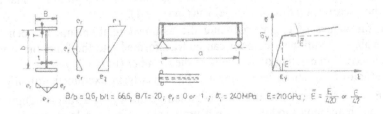

Figure 3.20: Models for the parametric study

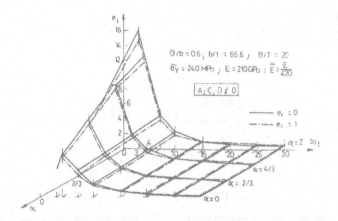

Figure 3.21: Compression fibre strains (e_1) characterizing the critical state of equilibrium for combined plate buckling and lateral-torsional buckling

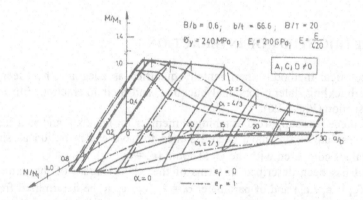

Figure 3.22: Transformed version of Fig. 3.21, illustrating combinations of normal forces and bending moments which belong to the critical state of equilibrium

specific strain values e_1 characterising the critical state have been sought for given values ($\alpha = 0$, $\alpha = 2/3$, $\alpha = 4/3$ and $\alpha = 2$) of the load parameter. Variations of e_1 values versus α and the span-to-depth ratio a/b are shown in Fig. 3.21 for negligible residual strain ($e_r = 0$) and for those corresponding to the yield strain ($e_r = 1$).

From the values of e_1 and α, using the assumed relationship between strains and stresses, the above mentioned diagram can be transformed into the function of the bending moment and axial force acting on the ends of the member (Fig. 3.22).

The effect of the strain hardening modulus has been examined utilising results for ratios $\overline{\overline{E}}/E = 1/420$ and $1/42$. For the assumed geometry, at a span-to-depth ratio $a/b = 15$, the strain hardening modulus did not practically affect the load capacity, but below this value it may be overwhelming.

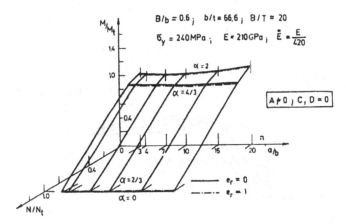

Figure 3.23: Characteristic normal force–bending moment combinations causing web plate buckling, when lateral-torsional buckling is prevented

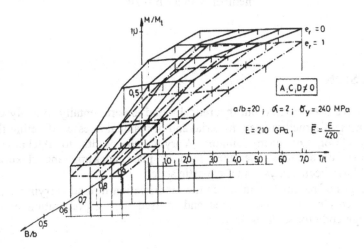

Figure 3.24: Influence of flange width to web depth ratio for the load carrying capacity of a bended member with a / b = 20

Residual strains have the greatest effect at about **a / b = 15**, in the so-called medium' slenderness range (with the assumed data, $\lambda = 87$).

Excluding the possibility of lateral-torsional buckling, the relationship in Fig. 3.23 is obtained, where, with the geometry assumed, the minimum of the load carrying capacity is at a ratio **a / b = 7** (assuming a lengthwise wave). Excluding the possibility of web buckling, a diagram rather similar to Fig. 3.22 results, with a deviation distinct near **a / b = 7**.

Concerning the effect of geometry on load carrying capacity, for a ratio **a / b = 20** (which is usually in construction and in pure bending) the ratio **B / b** of flange width: web depth was seen to affect markedly the load capacity (Fig. 3.24). The greatest load capacity decrease due to residual strains was found at about **B / b = 0.7** (Fig. 3.25).

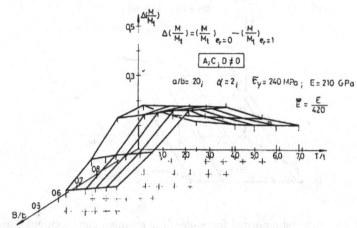

Figure 3.25: Load carrying capacity reduction effect of residual stresses for a bended
member with **a / b = 20**

3.7 CONCLUSION

Investigations have been carried out theoretically and experimentally to analyse the effect
of strain hardening of structural steel material and of residual stresses affecting the stability
(instability) conditions of beam-columns. Varying the width to thickness ratios of
component plates of sections, the interaction of plate buckling and flexural-torsional
buckling could have been analysed in the plastic state.

   Those ranges of parameters in which the modification in load carrying capacity is of
higher importance have been determined and utilised during preparation of the newest
Hungarian design code for steel construction.

CHAPTER FOUR

ULTIMATE LOAD BEHAVIOUR OF STEEL FRAMED STRUCTURES

4.1 INTRODUCTION

The increasingly powerful experimental and computational tools of structural design require well-defined design philosophy. As its basis the concept of limit states [84] is seemingly accepted in many countries, requiring the estimation of the (fairly small) risk that the given structure be brought to its ultimate state (failure) and the (somewhat bigger) risk of the occurrence of phenomena restricting its regular use (serviceability). All this (excluding now brittle fracture and fatigue) necessitates the analysis of structural response over a broad range of load levels, from working loads up to exceptionally high ones.

As pointed out in the literature [85], in different periods of developing engineering practice different importance was attributed to the two classes of limit states. In the earliest periods (e.g. in the works of Coulomb) - possibly inspired by the experience of collapsing vaults (Fig. 4.1) and breakdown of earthworks - interest was focused on the ultimate state, and accordingly the applied methods of analysis could describe only the last phase of structural response (it is still in use, e.g. in some branches of soil mechanics).

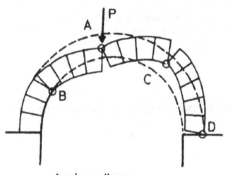

Arch collapses as a mechanism when hinges form at B, A, C and D

Figure 4.1: Possible mode of collapse of masonry arch

A second period can be connected to the activity of the brilliant French scientist Navier, who seems to have been more interested in the second class of limit states (Fig. 4.2). Quoting the preface of his 1826 book, which is of enormous influence on engineering practice [86]: "Knowing the cohesion, the ultimate load to be carried by a body can be determined. For the structural engineer, however, this is not sufficient, the question being not to know the force great enough to cause breakdown of the body, but rather the load to be carried by the structure without causing in it changes progressing with time". From this viewpoint originated the concept of allowable stresses and the corresponding methods of analysis: the "accurate" and simplified methods based on the theory of elasticity which were good enough to describe structural response at relatively low (working) load levels.

Let us associate a subsequent period with the work of the Hungarian scientist Kazinczy, who is regarded as the initiator of plastic design of steel structures. In his early - and because of its language hardly accessible - paper (1914) on his tests with fixed-end beams he confesses [87]: "... In the case of statically redundant steel structures exhibiting a different kind of response to higher loads than to lower ones ... the allowable stress is meaningless, giving no information whatsoever about the margin of safety". This indicates that main interest was shifting again towards the first class of limit states, towards failure.

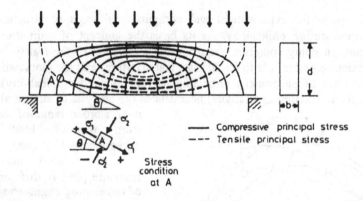

Figure 4.2: Principal stresses in a beam carrying a uniformly distributed load

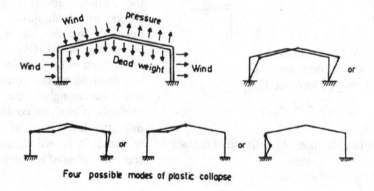

Figure 4.3: Four possible modes of collapse of a single-storey frame

Accordingly research was directed to complete the methods of analysis with new ones (based on the theory of plasticity among others) describing structural behaviour in the vicinity of and at the peak load, and often in post-failure phase as well (Fig. 4.3).

Thus the recent concept of limit states can be regarded as a balanced synthesis of the previous design philosophies.

4.2 DIFFICULTIES IN PREDICTING FAILURE

In contrast to the expectation of the initiators of plastic design, the analysis of structural response in the vicinity of peak load proved to be extremely complicated, due not only (and even not mainly) to inelastic behaviour, but to the fact that in the vicinity of peak load

- change in geometry (geometrical non-linearity) gains importance over other factors because the effect of initial geometrical imperfections (often negligible at lower load levels) is magnified;
- residual stresses (remaining latent at lower loads) interact with growing active stresses resulting in premature plastic zones; and last but not least
- the usual and widely accepted tools of analysis, e.g. beam theory based on the Bernoulli–Navier theorem or the small-deflection theory of plates, restricting the actual degree of freedom of the structure, cannot describe exactly enough its real response at failure.

These difficulties can be overcome in the case of simple structural elements (separated compression members, parts of plate girders, etc.) by using more refined (e.g. finite element) methods, allowing degrees of freedom (e.g. distortion of cross-sections) excluded in traditional analysis, or even in the case of statically non-redundant structures, where the above indicated complex behaviour is usually confined to a limited section of the whole structure. Then the procedure can be illustrated by Figs. 4.4 and 4.5.

Fig. 4.4 indicates the classical method: (1) finding the appropriate constitutive law; (2) using basic notions and equations of the mechanics of continua; (3) establishing a mathematically treatable simplified model of the structure using, for example, simple beam theory; (4) describing load history by means of a load trajectory in a load space; (5) computing the corresponding change of primary parameters (trajectory of load actions, stresses, deflections) describing structural response; (6) selecting from these the so-called quality parameters v_1 (e.g. maximum moments or stresses) playing a decisive role in judging the onset of limit states [88], which again can be defined by the intersection of the trajectory of quality parameters with a given limit surface.

If the simplified model is not elaborate enough to reflect real structural behaviour, a secondary, more detailed, local model is inserted (see Fig. 4.5) to depict the mostly critical part of the structure, by which more realistic quality parameters (and limit surface) can be deduced from the primary parameters already known.

Because of the interaction of local and global behaviour, this pattern cannot be followed in the case of hyper-static structures, as the additional information gained by the secondary local model is to be fed back to the computation of primary parameters as well. For this purpose, if - as happens very often - the secondary model can be analysed by numerical methods or only experimentally, the results have either to be re-interpreted to gain mathematically treatable, sufficiently simple rules, or the secondary model has to be

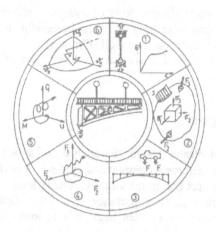
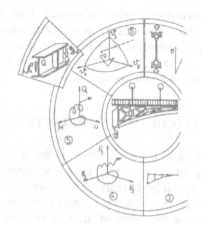

Figure 4.4: Classical method of analysis of structures   **Figure 4.5:** Secondary, more detailed, local model of analysis of structures

simplified to furnish digestible results. In both cases the validity or accuracy has to be proved by failure tests with full-scale structures (which are usually very expensive). The same applies to quantities which cannot be measured directly or which are hardly measurable, such as residual stresses.

4.3 TEST PROGRAMME

The experimental research project was carried out in the laboratory of the Department of Steel Structures, Technical University, Budapest.

Three main series of experiments were covered by the test programme.

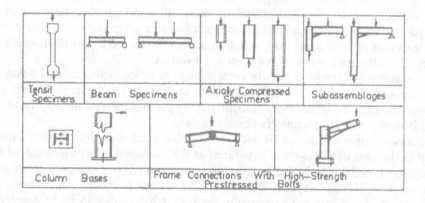

Figure 4.6: Test programme of additional tests on structural elements

The first part of the programme comprised the additional tests on stub columns, frame corners, plate elements and simple beams [89]. A brief summary of this series is given in Fig. 4.6.

In the second part of the programme the full-scale tests of frames have been included. Fig. 4.7 gives a brief summary of the full-scale tests and dimensions of the specimens, indicating the loads and the characteristics of the loading process. Test frame C-3 had rafters with a slope of 30% (16.7º), welded column I sections of 360 x 170 x 48, and welded rafter sections I 300 x 150 x 37. Rafter-to-column and mid-span connections were end-plated with high-strength preloaded bolts.

Different types of lateral supports were applied (Fig. 4.8) to the frames to prevent or decrease out-of-plane displacements and/or rotations of selected sections.

Vertical loads at the purlin supports were applied to the upper flange of rafters, so that webs and bottom flanges were not restricted laterally. To make horizontal displacements (sideways) unrestricted, the jacks were fastened not directly to the floor slab,

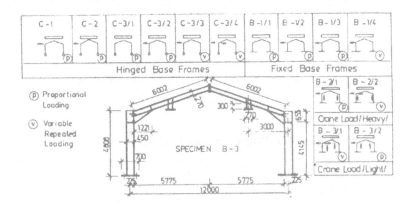

Figure 4.7: Test programme of the full-scale tests

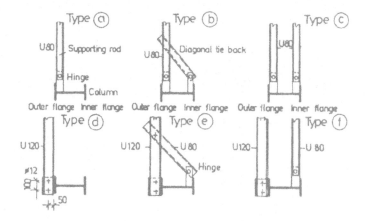

Figure 4.8: Different types of lateral supports

but through a so-called gravity load simulator (Fig. 4.9). The latter consisted of three elements: two bars and a rigid triangle. The two bars had pinned joints at both ends, resulting in a mechanism with one degree of freedom. Hydraulic jacks were joined to the rigid triangle. This mechanism produced a vertical load acting upon the intersection of the two bar axes. A characteristic curve of the simulator, illustrating the ratio of horizontal and vertical load components as a function of horizontal displacement, is given in Fig. 4.9.

The third part of the programme involved a representative part of a multipurpose, pinned, pitched roof industrial hall: a building section consisting of three frames, bracings with pinned elements, light gauge purlins and wall beams with corrugated steel sheeting.

Structural details of the building sections are shown in Fig. 4.10. The scope of

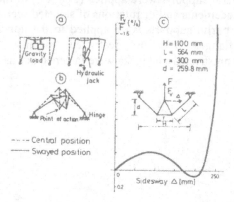

Figure 4.9: Role of gravity load simulator

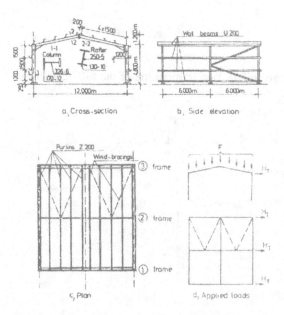

Figure 4.10: Structural details of the building sections

investigations was threefold:
- the effect of the restraint system on elastic behaviour,
- the residual deformation and load bearing capacity due to cyclically repeated loading,
- ultimate load of the frames.

Measurements in the elastic range of behaviour were made at the six different stages of erection.

Non-elastic tests (cyclically repeated load, incremental collapse) were carried out on the building section corresponding to state (f) using a load combination composed of dead and meteorological loads.

4.4   COMPUTATIONAL MODELS

The traditional concept of plastic design of steel structures is based on the assumption that, under gradually increasing static loads, plastic zones develop and grow in size and number, and eventually cause unrestricted, increasing deflections at the onset of the ultimate limit state of the structure. The concept was first introduced by Kazinczy [87] by establishing the concept of plastic hinge. Some basic questions are still discussed, among which are the effects of the difference between the ideally plastic constitutive law and the actual behaviour of steel material, and the consequence of local instability (plate buckling, lateral buckling). Joining in the international research in this field, we tried to introduce the concept of interactive plastic hinge, which can be substituted for the classic concept of the plastic hinge in the traditional methods of limit design, but can reflect the effect of phenomena such as strain hardening, residual stresses, plate buckling and lateral buckling [91].

4.5   INVESTIGATION OF PLATE BUCKLING WITH THE AID OF YIELD MECHANISMS

In the course of plate experiments, if the thickness to width ratio is small, the plate does not lose its load bearing capacity with the development of plastic deformation, but is able to take the load causing yield until a deformation characteristic of the plate occurs; it is even able to take a small increase in load. In the course of the process "crumpling" (buckling) can be observed on the plate surface. These "crumplings" are formed by a yield mechanism, with the plastic moments acting in the linear plastic hinges (peaks of waves) not constant but ever-increasing due to strain hardening. The yield mechanism performed by "crumpling" extends to the component plates of the bar. The description of its behaviour is obtained, from among the extreme-value theorems of plasticity, with the aid of the theorem of kinematics.

Thus, in the course of our investigations, an upper limit of load bearing capacity has been determined. However, to be able to assess the results, the following have to be considered: on one hand, the yield mechanisms are taken into account through the "crumpling" forms determined experimentally; and on the other hand, the results of theoretical investigations are compared with the experimental ones.

4.51 Yield mechanism forms based on experimental results

The different forms of yield mechanisms can be determined on the basis of experimental results. The yield mechanism forms of an I section bar can be classified according to the following criteria.
(a)   According to the manner of loading.
(b)   According to the positions of the intersecting lines of the web and the flanges, the so-called throat lines; thus
   (i)   the evolving formation is called a planar yield mechanism if the two throat-lines are in the same plane after the development of the yield mechanism.
   (ii) the evolving formation is called a spatial yield mechanism if the two throat-lines are not in the same plane after the development of the yield mechanism.

4.511 Yield mechanism of compression members

(a) Planar yield mechanism. The buckled form of the compression member and the formation of the chosen yield mechanism are shown in Fig. 4.11. Plastic deformation occurs in the hatched regions.

As an effect of compressive force N, compression develops. The yield mechanism is denoted by **(N)$_P$** where **P** stands for the planar yield mechanism formation.

(b) Spatial yield mechanism. The formation of the spatial yield mechanism for a compression member is shown in Fig. 4.12. The ends of the member are assumed to be hinge supported in both main inertia directions. The yield mechanism that has occurred is called spatial; the phenomenon models the planar buckling of the compression member or the buckling of the component plates in the course of buckling.

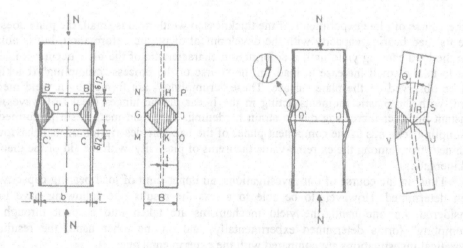

Figure 4.11: Planar yield mechanism of   **Figure 4.12:** Spatial yield mechanism of
                    compression member   compression member

The symbol of the yield mechanism is $(N)_S$ where S stands for the spatial yield mechanism formation.

4.512 Yield mechanisms of bent members

4.512.1 Bending moment constant along the member axis.
(a) **Planar yield mechanism.** The buckled form of the bent specimen and the chosen yield mechanism formation are shown in Fig. 4.13. As an effect of moment M, a rotation Θ develops.

As an effect of **M**, tension and compression regions develop.

The symbol of the yield mechanism is $(MC)_P$, where **C** stands for the constant bending moment.

(b) **Spatial yield mechanism.** The form of the spatial yield mechanism in the case of a bent rod is indicated in Fig. 4.14. The rod ends are assumed to be hinge supported in both

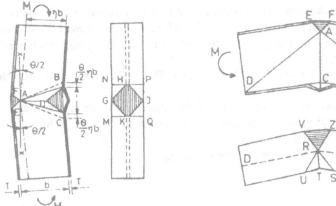

Figure 4.13: Planar yield mechanism of bent member with bending moment constant along the member axis

Figure 4.14: Spatial yield mechanism of bent member with bending moment constant along the member axis

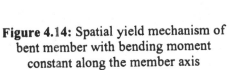

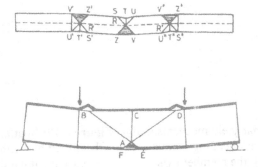

Figure 4.15: Yield mechanism of a simply supported beam specimen

main inertia directions. The yield mechanism models the buckling of the component plates of the bent member, the lateral buckling of the beams as well as their interaction.

The buckled formation of a simply supported beam specimen and the chosen yield mechanism are shown in Fig. 4.15.

In case of the yield mechanism formations in Fig. 4.15, the effect of neighbouring supports (the effect of ribs) has also been taken into account. As an effect of the moment, a rotation Θ develops.

The symbol of this yield mechanism is $(MC)_S$.

4.512.2 Bending moment varying along the rod axis.

In the case of a bending moment varying along the member axis, it is assumed that the "crack" of the web plate of the I section in the cross-section of the concentrated force is hindered by the thickness of the web plate or by the ribs.

Climenhaga and Johnson [92] assumed yield mechanism forms similar to those introduced in the preceding paragraph for the investigation of buckling occurring in the steel beam part of a composite steel–concrete construction.

(a) Planar yield mechanism. The buckled form of a bent specimen and the selected yield mechanism are shown in Fig. 4.16. As an effect of the moment, a rotation Θ develops.

Because of the clamping of the cross-section EC, the yield mechanism loses its symmetric character.

The symbol of the yield mechanism is $(MV)_P$ where V stands for the varying moment.

(b) Spatial yield mechanism. The form of the spatial yield mechanism in the case of a bending moment varying along the rod axis is shown in Fig. 4.17. As an effect of the moment, a rotation Θ develops.

The symbol of the yield mechanism is $(MV)_S$.

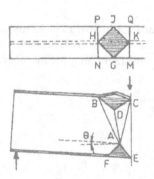

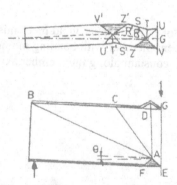

Figure 4.16: Planar yield mechanism of bent member with bending moment varying along the member axis

Figure 4.17: Spatial yield mechanism of bent member with bending moment varying along the member axis

4.513 Yield mechanism of the component plates of an I section member

Yield mechanism formations have been determined for different stresses. On the basis of the experimental results it is expedient to decompose these yield mechanism formations into the yield mechanism formations of the component plates of an I section rod, as certain component plate formations appear in other yield mechanisms too.

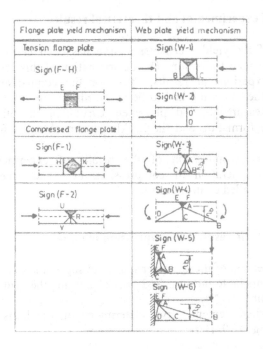

Figure 4.18: Yield mechanisms of the component plates of an I section member

To classify the yield mechanisms of component plates, the following classification has been used.

(a) Flange plate, if the plate is supported along the central line.
(b) Web plate, if the plate is supported at the unloaded ends.

Fig. 4.18 shows the yield mechanisms of the component plates where **F** is the flange plate, **W** is the web plate; the odd numbers refer to the planar yield mechanisms and the even ones to the spatial yield mechanisms.

4.52 "Joining" the yield mechanisms of component plates

The "joining" of the yield mechanisms of component plates depends on the positions of the so-called throat-lines of the yield mechanism chosen on the basis of the experimental results.

In cases pertaining to planar yield mechanisms, this "joining" is to be realised in a linear manner, with a linear plastic hinge: the length of the linear plastic hinge is governed - due to the properties of the chosen yield mechanism - by the length of the yield mechanism of the compression flange plate (F-1). In the case of spatial yield mechanisms, the "joining" should be realised at one or more points.

4..53 Comparison of experimental and theoretical results

The determination of the yield mechanism curve has been carried out by assuming a particular yield mechanism formation. The determination of this formation was primarily made possible by experimental results; thus, for the comparison of theoretical and experimental results the following had to be analysed: on one hand, the relation between the computed and measured load–displacement curves; on the other hand, the relation between the assumed formation and the buckling form obtained from the experimental results.

The basis of the comparison of the results of experimental and theoretical investigations was consideration of compression, bent and eccentrically compressed members as well as so-called control beams.

4.531 Investigation of the results of compression members

Because of the arrangement of the specimens, the end support is clamped from the point of view of bending and torsion to the so-called "weak" axis and the evolving yield mechanism is a planar one; its symbol is $(N)_P$.

The four load–displacement curves determined on the basis of the experimental results are shown in Fig. 4.19. Here:

$$N_p = 2B \cdot T \cdot \sigma_{YF} + b \cdot t \cdot \sigma_{YW}$$

$$\delta_P = \frac{N_P \cdot L}{EA}$$

The relation and coincidence of the experimental and theoretical results is acceptable.

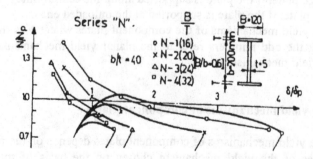

Figure 4.19: Load–displacement curves of compression members

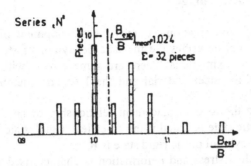

Figure 4.20: Measured buckling wave length to the width of the flange plate of compression members

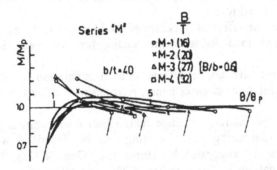

Figure 4.21: Load–displacement curves of bent beams

Figure 4.20 indicates the results of measurement of forms evolving as an effect of plate buckling that served to disclose the form of buckling of the flange plate. The buckling of the component plates are in interaction with each other and thus it is sufficient to investigate the buckling formation of the flange plate. The assumed buckling form of the flange plates has a length **B** (Fig. 4.21); the diagram compares the measured buckling wavelengths B_{exp} with the width **B** of the flange plate. It can be observed that the assumed yield mechanism formation coincides with that obtained from the experimental results.

4.532 Investigation of the results of bent specimens

The buckling investigation of component plates of bent beams can also be carried out with the aid of the yield mechanism curve.

Because of the arrangement of the specimens, the evolving yield mechanism is a planar one; its symbol is $(M)_P$.

The load–displacement curves determined on the basis of the experimental results fit the curve determined with the aid of the yield mechanism (see Fig. 4.21). Here:

$$\Theta_P = \frac{M_P \cdot L}{2EI_x}$$

4.54 Model of the interactive hinge

The plastic load bearing investigation assumes the development of rigid–ideally plastic hinges; however, the model describes the inelastic behaviour of steel structures but with major constraints and approximations. There are some effects with the consideration of which the behaviour of the steel material and the I section member can be taken into account more realistically.

(1)   When determining the load–displacement relationship of an I section member, the symbol of the elastic state is **E** and if the so-called "rigid" state is assumed instead of the elastic one, the symbol of the rigid state is **R**.

(2)   The effect of residual stress and deformation is characterised by a straight line for ease of handling. The symbol used when taking the residual stress and deformation into consideration is **O**.

(3)   Strain hardening is one of the important features of the steel material; **Ŝ** indicates that it has been accounted for.

(4)   In Section 4.52, the effect of buckling of the I section member component plates on the rod element load–displacement relationship has been investigated; this is indicated by **L**.

The models that take the above effects into consideration in the investigation of load–displacement (relative displacement) relationships of an I section are called "interactive" ones.

The model of the interactive hinge taking into consideration the effect of rigid–residual stress–strain-hardening–plate buckling can be described with the aid of the "equivalent beam length" suggested by Horne [96] (Fig. 4.22a). The material model applied in the investigations is shown in Fig. 4.22b. The effect of the residual stresses and deformations is substituted by a straight line. The effect of strain hardening can be determined with the help of the rigid–hardening (**R-S**) model. The buckling of the I section

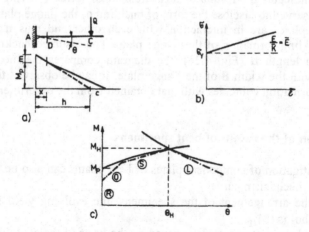

Figure 4.22: Model of interactive hinge

member component plates is described by the yield mechanism curve, which is substituted by a straight line.

Figure 4.22c indicates the load–displacement relationship of the (R-O-S-L) interactive hinge. The substitution by straight lines is justified to simplify the investigations. In the (R-O-S) sections the intersections are connected while in section L the moment–rotation relationship is substituted by a tangent that can be drawn at the apex.

4.6 COMPUTATIONAL MODEL FOR ULTIMATE LOAD OF STEEL FRAMES

The aim is a theoretical and experimental investigation that studies steel structures in steps known from traditional methods with the aid of a hinge model suitable to describe more "refined" properties, to embrace more phenomena (strain hardening, residual deformation, plate buckling, flexural-torsional buckling, etc.).

4.61 Effect of local instability

The final collapse of steel structures is mostly caused by instability phenomena [97], which may be:

    (a)   disadvantageous change in the steel structure geometry,

    (b)   disadvantageous change in the cross-section geometry.

The effect of (a) can be traditionally grouped in the field of plastic instability. Halász [98] treated the problem in his doctoral thesis and, over and above theoretical studies, he also introduced a method suitable for practical design work.

The disadvantageous changes in cross-section geometry are mainly plate buckling. Plate buckling causes a change in the behaviour of the plastic hinge, too, and thus, for the plastic investigation of statically indeterminate steel structures, methods have to be elaborated that take the effect of plate buckling into consideration too.

The material should be assumed to be ideally plastic while investigating the effects of the evolving plastic hinges, lest the previously formed plastic hinges should "close".

Studying the effect of strain hardening and plate buckling, one should bear in mind that the load–displacement diagram of the structure may be of an ascending type even if the characteristic curves of the given member section or sections are of a descending type in individual cross-sections because of the effect of plate buckling.

In the theory of plasticity, when deriving the condition of plasticity or some other physical relationships, Drucker's postulate for stability is applied, derived by assuming stable materials [99].

It should be noted that Drucker's postulate is not a natural law but a criterion of classification [100] – the materials very often do not correspond to the assumptions of stable materials, or structural elements may behave unstably, while at the same time their material is in a stable state.

Maier [101] was the first to treat the problem of the effect of the unstable state of certain members on the behaviour of a triangulated structure. Again it was Maier who in 1966 reintroduced the subject and investigated a structure consisting of compressed members and rigid beams where the load–displacement diagram of individual members contained stable and unstable parts.

Maier and Drucker [102] re-examined the original Drucker postulate applied when determining the condition of plasticity since the original postulate is suitable for the determination of the convexity and normality of the condition of plasticity in the case of stable materials only.

When studying the load bearing capacity of steel structures, the problem of unstable or softening material, according to Drucker's postulate, does not appear since the strain hardening of the steel material may increase to a major extent the plastic load bearing capacity of the steel structure. However, as it has been known for a long time, the final collapse of steel structures is caused, in a high percentage of cases, by instability (plate buckling, flexural-torsional buckling) phenomena that may occur in the cross-section or in a structural unit.

Concerning steel structures, the properties of plastic hinges over and above the usual elastic–ideally plastic–hardening behaviour may be complemented with the effect of instability (flexural-torsional buckling) developing in the given structural unit (the environment of the plastic hinge).

This type of inelastic or interactive hinge describes the behaviour of the structural unit and at the same time also satisfies the criteria for an unstable or softening structural unit, according to the Maier–Drucker postulate.

When determining the plastic load bearing capacity of steel structures, the interactive hinge of softening character has so far not been considered or applied. The effects of the stability phenomena causing the softening character (flexural-torsional buckling, plate buckling) can be taken into account indirectly with the aid of construction rules. In principle, mathematical programming allows the investigation of more complex steel structures, too; however, it is less suitable for designing practice. The author [103] has suggested a procedure that takes into account the softening character of the inelastic hinge in the form of an interactive zone. The softening character of the interactive zone is caused by the buckling of the component plates, a phenomenon that can be studied with the help of the yield mechanism.

4.62 Computer program for steel frames

Recent development of engineering structures has made it necessary to increase the accuracy of existing computing methods.

4.63 Fundamentals of bar system computation

Matrix methods are available for computer determination of stresses in plane bar systems [105]. These methods rely either on the force or on the displacement method; this latter has been applied in the program.

A method well-known and proven in the specialist literature has been followed. As a first step, the structure is decomposed into nodes and rectilinear bars. Bar ends have three degrees of freedom in displacement: two displacements in the structure plane (u_x and u_y)

and rotation in this plane (ϕ_z). Accordingly, in any bar, three different stresses can be determined, i.e., due to normal force **N**, shear **T**, and bending moment **M**. In the knowledge of the bar stiffness matrices $\hat{\underline{K}}$, the overall stiffness matrix $\underline{K}$ of the structure can be compiled. Also, to reckon with supporting conditions, the known procedure was followed: cancelling from $\underline{K}$ block rows and columns corresponding to support numerals, thus computing a stiffness matrix of reduced size. For a given load, also, load vector $\underline{q}$ is known; hence, after solving the equilibrium equation system

$$\underline{K} \cdot \underline{u} = \underline{q} \quad \text{or} \quad \underline{u} = \underline{K}^{-1} \cdot \underline{q} \tag{4.1}$$

reaction forces are obtained from the system of nodal displacements. To increase the accuracy of description of the geometry, the second-order theory has been applied. Second-order computation for bars loaded by normal forces may apply the known stability functions [106]. As a consequence, bar stiffness matrix $\underline{K}$ will not be constant (as it is in the first-order theory), but it will be a function of the bar force parameter N / N_E (where N_E is the Euler critical load).

Thereby all the computation becomes iterative; solutions $\underline{u}_1$, $\underline{u}_2$, ... , $\underline{u}_n$ of the equilibrium equation system (4.1) converge after a number of computation cycles to the second-order solution.

4.631 The applied bar element

Simpler cases involve the bar element in Fig. 4.23, permitting fast, easy computations, usually on an elastic material model. The bar stiffness matrix $\hat{\underline{K}}$ is common knowledge; stiffness values are obtained by solving basic problems of hyper-static beams.

Our goal seemed to be better achieved by considering the complex bar element shown in Fig. 4.24.

Two end parts of the bar, of lengths l_1 and l_2, are infinitely rigid (maybe $l_1 = l_2 = 0$), the middle part is elastic. The rigid and elastic parts of the bar are connected by rotation springs, each of which is able to rotate through ϕ in the structure plane alone. Stiffnesses, i.e. spring constants, are c_1 and c_2.

This assumption of the bar element had the following motivation.

(1)   A plastic second-order analysis was the goal. Applying the usual bar

Figure 4.28: Simple bar element

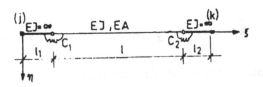

Figure 4.29: Complex bar element

elements, development of plastic hinges would require intercalation of a spring of stiffness **c** instead of a plastic hinge, each hence to increase the stiffness matrix $\underline{K}$ of the structure. The assumption of the bar element avoids this computational problem, by permitting simulation of the development of plastic hinges by changing the stiffness of the inner spring.

(2)   It provides the assumption of an ideal clamping device or ideal hinge exact enough for more accurate computations. This bar element also permits dealing with elastic clamping.

(3)   Structural design may produce infinitely stiff bar parts (gussets). Although it can be dealt with by assuming a value of **EI** sufficiently high for conventional bar elements, the stiffness matrix $\underline{\hat{K}}$ thereby becomes less appropriate, sometimes causing numerical errors. The bar element as presented helps to avoid this problem.

The stiffness matrix $\underline{\hat{K}}$ of the bar element can be written in the knowledge of bar reaction forces resulting from unit nodal displacements.

4.632 Spring characteristics

Spring characteristics are of the general form shown in Fig. 4.25. Sections have different spring constants $c = \Delta M / \Delta \Theta$ indicating the given section of the elasto-plastic behaviour or of the stability condition of the bar part. The characteristic is strictly monotonous for Θ but not for **M**, for which the curve has a peak followed by a descending path.

Application of spring characteristics lessens the validity of the theorem that a hyperstatic beam with **n** redundancies fails at the development of the $(n + 1)$th plastic hinge.

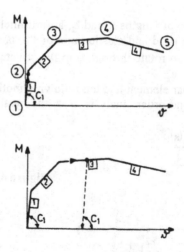

Figure 4.25: Spring characteristics of complex bar element

It is true only for computations relying on rigid-plastic behaviour - hence not for strain-hardening models. It is both logically obvious and demonstrated by computations that plastic reserves after yield permit more **n + 1** interactive hinges to develop. Final failure depends on a complex interaction between structural design, loss of stability phenomena and yield mechanism.

4.7 RESULTS OF THEORETICAL AND EXPERIMENTAL INVESTIGATIONS

Concerning the experimental frame C-3/2, the load–deflection curve develops according to Fig. 4.26. On the side of horizontal load, the first inelastic hinge develops due to the residual stresses and deformations in the cross-section beneath the frame knee, and this hinge develops at 52% of the maximal frame load. At 97% of the maximal load, Zone **L**, describing the effect of plate buckling, also develops in this cross-section, i.e. in the frame cross-section an "unstable" state - a descending characteristic curve develops.

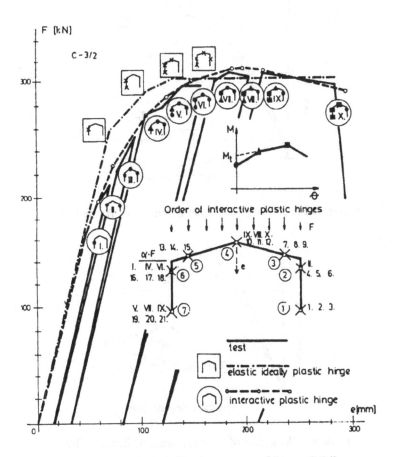

Figure 4.26: Load–deflection curves of frame C-3/2

For comparison, Fig. 4.26 introduces the characteristic load–displacement curve of the frame structure in the case, too, when the basis of the computations is the traditional plastic hinge.

4.71 Effect of fabrication and erection

The effect of incorrect geometry was investigated by introducing different initial lateral displacements. The consequences are illustrated in Fig. 4.27.

The effect of residual stresses of different intensities is shown in Fig. 4.28. The medium curve coincided with test results.

The method presented for the complex analysis of frameworks takes several effects into consideration (Fig. 4.29).

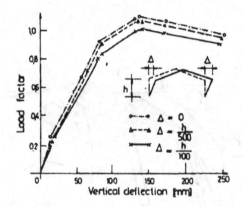

Figure 4.27: Effect of incorrect geometry of frame C-3/2

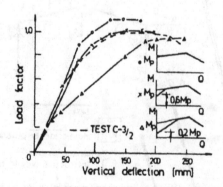

Figure 4.28: Effect of residual stress of frame C-3/2

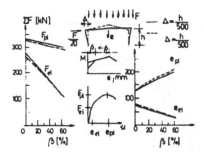

Figure 4.29: Complex analysis of frames

4.72 Effect of structural details

It seems worthwhile to draw attention to the occasional decisive effect of minor differences in structural details on failure as well. Some of the results are reproduced below.

4.721 Column bases

Column bases were fixed or hinged. The hinges were not ideal: columns could have been supported by larger base plates: Fig. 4.30 compares the measured bending moment due to vertical and horizontal load with the calculated ones assuming a pinned (broken line) and fixed (solid line) frame. The corresponding moment–rotation diagram of the column base was checked experimentally, its adaptation to an interactive plastic hinge indicating the load–deflection diagrams obtained by different end conditions (Fig. 4.31).

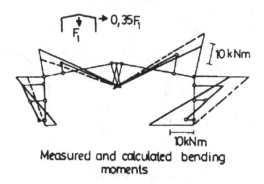

Measured and calculated bending moments

Figure 4.30: Measured and calculated bending moments

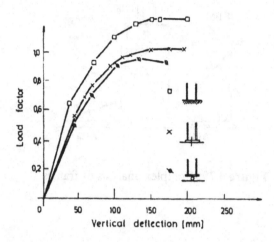

Figure 4.31: Effect of column bases in frames

4.722 Lateral supports

Spacing and efficiency of lateral supports proved to be of basic importance. Their effect is illustrated in Fig. 4.32.

The importance of adequate spacing of lateral supports and their efficiency in preventing the rotation of the cross-section around the bar axis has to be emphasised as purlins and rails connected to tension flanges often cannot be regarded as fully effective in the case of thin webs. Not only can the load carrying capacity thus be substantially reduced

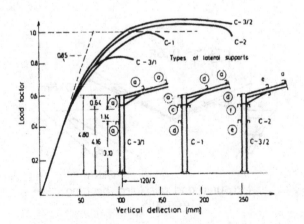

Fig 4.32: Importance of lateral supports in frames

(as by elastic lateral buckling in the case of frame C-3/1 in Fig. 4.32), but the yield plateau in the load–deflection diagram can be too short (as in the case of frame C-1 in Fig. 4.32), rendering the structure sensitive to initial imperfections.

4.74 Building under proportional loading

4.741 Cross-bracing of the end-frame

Measured deflections from uniform horizontal loads are shown in Fig. 4.33, representing the effect of both semi-rigid cross-bracings of the end frame.

4.742 Horizontal and vertical bracing system

Load–displacement diagrams of incremental collapse tests are shown in Fig. 4.34. Ultimate loads are influenced by local loss of stability, previous loadings and the layout of frame (horizontal and vertical bracing connections).

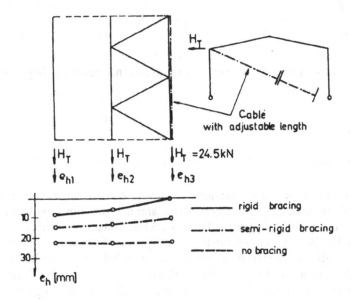

Figure 4.33: Effect of semi-rigid cross-bracings of the end frame

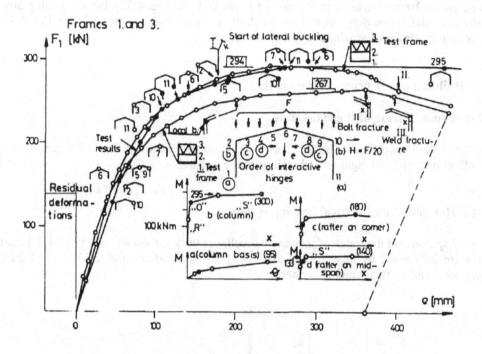

Figure 4.34: Load–displacement diagrams of incremental collapse tests

4.7 CONCLUSIONS

Experimental and theoretical investigations have been carried out in connection with the elastic and the plastic load bearing study of frame and hall structures, with strain hardening of steel material, the residual stresses and plate buckling taken into account.

A method has been presented for the investigation of frame structures applying the steps of known, traditional methods so that the structure behaviour can be analysed during the entire loading process. Certain effects determining the structural behaviour (e.g. residual deformation, steel material strain hardening and plate buckling) have been taken into consideration with the aid of the interactive plastic hinge concept. The interactive hinge was incorporated into an investigation method operating with the structure matrix-calculation method. The results of the elaborated method have been compared with the experimental investigation of full-scale structures.

REFERENCES

Chapter One

1. MASSONNET, ČH.: Design practices in Europe, in: Proceedings of the Int. Coll. on Stability Structures, Washington, 1977, 503-531.
2. HALÁSZ, O.: Stability problems in national specifications, in: Final Report to the Reg. Coll. on Stability of Steel Structures, Budapest, 1977, 9-13.
3. BOLOTIN, V. V.: Статистические методы в строителңой механике, Izd. Lit. Stroit., Moscow, 1965.
4. HALÁSZ, O. and PLATTHY, P.: A számítási modellek megbízhatóságának néhány elvi kérdése acélszerkezetek esetében, Mélyépítéstudományi Szemle, XXIV (1974), 89-95
5. KOITER, W. T.: Over de stabiliteit van het elastisch evenwicht, Thesis, Delft, 1945.
6. THOMPSON, J. M. T. and HUNT, G. W.: A general theory of elastic stability, John Wiley and Sons, 1973.
7. SKALOUD, M.: General report on plate and box girders, Final Report of Reg. Coll. of Steel Structures, Budapest, 1978.
8. KÁRMÁN, T.: Festigkeitsprobleme im Maschinenbau, Enzykl. der Math. Wiss., 1910., IV. 348-351.
9. MARGUERRE, K.: Die mittragende Breite des gedrükten Plattenstreifens. Luftfahrtforschung 14., 1937.
10. MAQUOI, R. and MASSONNET, CH.: Theorie non-linearie de la résistance postcritique des brandes poutres au caissons raidies, IABSE Publ., Zürich, 1971, 3-II.
11. MERISSON ET AL.: Inquiry into the basis of design and method of erection of steel box girder, Report, London, 1974.
12. AUGUSTI, G.: Stabilita di strutture elastiche elementari in presenza di grandi spostamenti, Atti Acc. Sci. fis. mat., Napoli, 1964, 3, 4, 5.
13. KOITER, W. T. and PIGNATARO, M.: A general theory for the interaction between overall and local buckling of stiffened panels, Rep. 556, Delft, Univ., 1976.
14. VAN DER NEUT, A.: The interaction of local buckling and column failure of thin-walled compression members, Proc. 12th Int. Congr. Appl. Mech., Stanford Univ., (1968), Springer-Verlag, 1969
15. DOWLING, P. J. and CHATTERJEE, S.: Design of box girder compression flanges. Design of webs of plate and box girders, 2nd Int. Coll. Stability of Steel Str., Introductory Report, Liège, 1977, 153-208.
16. KÁRMÁN, T., SECHLER, E. E. and DONELL, L. H.: The strength of thin plates in compression. Trans. Am. Soc. Mech. Eng., 54 (1932).
17. WAGNER, H.: Ebene Blechwandträger mit sehr dünnem Stegblech, Zeitschr. f. Flugtechnik und Motorluftschiffart, 20 (1922).
18. KLÖPPEL, K., SCHMIED, R. and SCHUBERT, J.: Die Traglest mittig und außermittig gedrückter dünnwandigen Stützen, Der Stahlbau, Nr. 1.3 (1969).
19. MAQUOI, R. and MASSONNET, CH.: Interaction between local plate buckling and overall buckling in thinwalled compression members, IUTAM Symp. Cambridge, Springer, 1976.

20. SKALOUD, M.: The limiting state of thin-walled columns with regard to the interaction of the deformation of the column as a whole with the buckling of their plate elements, Acta Technica CSAV, No. 8 (1967).

21. SKALOUD, M. and ZÖRNEROVÁ, M.: Experimentálni vysetrováni interakce vzperu tenkosténnych prutu a bouleni jejich sten, Final Res. Report Czechoslovak Acad. Sci. Inst. of Theor. Appl. Mech., Prague, 1969.

22. SKALOUD, M. and ZÖRNEROVÁ, M.: Experimental investigation into the interaction of buckling of compressed thin-walled columns with the buckling of their plate elements, Acta Technica CSAV, No. 4 (1970).

23. SKALOUD, M. and NAPRSTEK, J.: Limit state of compressed thin-walled steel columns with regard to the interaction between column and plate buckling, in: 2nd Int. Coll. on Stability of Steel Structures, Liège, 1977, 405-414.

24. REIS, A. J. and ROORDA, J.: The interaction between lateral-torsional and local plate buckling in thin-walled beams, in: 2nd Int. Coll. on Stability of Steel Structures, Liège, 1977, 415-427.

25. BASLER, K.: Strength of plate girders in shear, ASCE St. 7, 1961, 151-180.

26. BERGMANN, S.: Behaviour of buckled rectangular plates, Doctoral Thesis, Stockholm, 1948.

27. SKALOUD, M.: Interaktion der Ausbeulung von Wänden und der gesamten Formänderung gedrückter und gebogener Stäbe, Acta Technica CSAV, No. 1 (1962).

28. SKALOUD, M.: Effet d'une courbure initiale sur le compostement post-critique, Acier 5 (1965), 249-254.

29. SKALOUD, M.: Postbuckled behaviour of stiffened webs, Transactions of CSAV, 80 (1970), 1-154.

30. SKALOUD, M. and NOVAK, P.: Post-buckled behaviour of webs under partial-edge loading, Transactions of CSAV, 85(3) (1975) 1-94.

31. SKALOUD, M. and ZÖRNEROVÁ, M.: Optimum rigidity of longitudinal stiffeners of compression flanges, Acta Technica CSAV, No. 5, (1976), 549-580.

32. MASSONNET ET AL.: Plate and box girders, in: Introductory Report 2nd Int. Coll. on Stability of Steel Str., Liège, 1977, 145-208.

33. ROCKEY, K. C. and SKALOUD, M.: The ultimate load behaviour of plate girders in shear, The Structural Engineer, 50(1) (1972).

34a. IVÁNYI, M.: Yield mechanism curves for local buckling of axially compressed members, Periodica Politechnica (Budapest), Civ. Eng., 23(3-4) (1979), 203-216.

34b. IVÁNYI, M.: Moment–rotation characteristics of locally buckling beams, Periodica Politechnica (Budapest), Civ. Eng., 23(3-4) (1979), 217-230.

35. CROLL, J. G. A. and WALKER, A. C.: The Elements of Structural Stability, Macmillan, London, 1972.

Chapter Two

36. BRYAN, G. H.: On the stability of a plane plate under thrust in its own plate with application to the buckling of the side of a ship, Proc. London Math. Society, 22, 1891.

37. TIMOSHENKO, S.: Sur la stabilité des systemes élastiques, Ann. des Ponts et chauss., 15, 1913.

38. REISSNER, H.: Über die knicksichermeit ebener bleche, Zentralblatt der Bauverwaltung, 1909.

39. SEZAWA, K.: Stability of a thin plate, Journal of the Society of Naval Architecture, Japan, 38 (1925).

40. WAGNER, H.: Über konstruktions und Berechnung-fragen des Blechbaues, Jb. Wiss., Ges. Luftf., 1928.

41. TAYLOR, G. I.: The buckling load for a rectangular plate with four clamped edges, Z. Angew. Math. Mech., 13 (1933).

42. BLEICH, F.: Theory und Berechnung der eiseren Brücken, Julues Springer, Berlin, 1924.

43. CHWALLA, E.: Reports 2nd International Congress, Brodge and Structural Engineer, Wien, 1928.

44. ROS, M. and EICHINGER, A.: Reports 3rd International Congress, Bridge and Structural Engineer, Paris, 1932.

45. HENCKY, H.: Zur Theorie plasticher Deformationen und der Heirdrich im Material Hervorgerofenen Nebenspannungen, Proc. 1st International Congress, Appl. Mech., Delft, 1924.

46. PRANDTL, L.: Spannungsverteilung in plastischen Koerpern, Proc. 1st International Congress, Appl. Mech., Delft, 1924.

47. REUSS, E.: Berucksichtigung der elastischen Formaenderungen in der Plastizitats Theorie, 2. Angew. Math. Mech., 10 (1930).

48. BIJLAARD, P. P.: Theory of the plastic buckling of thin plates, IABSE, Zürich, 6 (1940-41).

49. ILYUSHIN, A. A.: Some Problems in the Theory of Plastic Deformations, Prikladnaya Matematika i Mechanika, 7 (1943), p. 245. Translation by Appl. Math. Group, Brown Univ., 1946.

50. STOWELL, E. Z.: A unified theory of plastic buckling of columns and plates, NACA Technical Note No. 1556, 1948.

51. STOWELL, E. Z. and PRIDE, R. A.: The effect of compressibility of the material in plastic buckling. J. Aero. Sci., 18 (1951).

52. SHANLEY, F. R.: Inelastic column theory, J. Aero. Sci., 14 (1947).

53. HANDELMANN, C. H. and PRAGER, W.: Plastic buckling of a rectangular plate under edge thrusts, NACA Technical Note No. 1530, 1948.

54. PEARSON, C. E.: Bifurcation criterion and plastic buckling of plates and columns, J. Aero. Sci., 17 (1950).

55. ONAT, E. T. and DROCKER, D. C.: Inelastic instability and invremental theory of plasticity, J. Aero. Sci., 20 (1953).

56. YAMAMOTO, Y.: A general theory of plastic buckling of plates, J. Soc. Naval Arch. (Japan), 96 (1955).

57. THURLIMANN, B. and HAAIJER, G.: On inelastic buckling in steel, Proc. ASCE, April 1958.

58. BOULTON, N. S. and LANCE MARTIN, H. E.: Residual stresses in arc-welded plates, Proc. Inst. Mech. Eng., London, 123, 1936.

59. ROSENTHAL, D.: Mathematical theory of heat distribution during welding and cutting, Welding Journal Res. Suppl., 20 (1941).

60. FUJITA, Y.: Buit-up column strength, Ph.D. Diss., Lehigh Univ., August 1956, Univ. Microfilms, Michigan.
61. TALL, L.: The strength of welded built-up columns, Ph.D. Diss., Lehigh Univ., May 1961, Univ. Microfilms, Michigan.
62. GRIFFITHS, G. H.: Residual stresses in butt-welded steel plates, Welding Journal Res. Suppl., 20 (1941).
63. PATTON, E. O., GORBOUNOV, B. M. and BERSHTEIN, D. O.: Behavior of residual stresses under external laod and their effect of strength of welded structures (translated by CORDOVI, M. A.), Welding Journal Res. Suppl., 23 (1944).
64. OKERBLOM, N. O.: Расчет деформации металлоконструкции при сварке, Mushgiz, Moscow, 1955.
65. OKERBLOM, N. O.: The influence of residual stresses on stability of welded structures and structural members, Commission X, Int. Inst. Welding, Liège, 1960.
66. YOSHIKI, M., FUJITA, Y. and KAWAI, T.: Influence of residual stresses on the buckling of plates, J. Soc. Naval Arch. (Japan), 107 (1960).
67. SOKOLNIKOFF, I. S.: Mathematical theory of elasticity, 2nd ed., McGraw-Hill, New York, 1956.

Chapter Three
68. KALISZKY, S.: Plasticity: Theory and Engineering Applications, Akadémiai Kiadó, Budapest, 1989.
69. KACHANOV, L. M.: Foundations of the Theory of Plasticity, North-Holland, Amsterdam–London,1971.
70. PEITER, A.: Eigenspannungen I. Art, Michael Triltsch Verlag, Düsseldorf, 1966.
71. CROLL, I. G. A. and WALKER, A. C.: Elements of Structural Stability, Wiley, London, 1972.
72. THOMPSON, J. M. T. and HUNT, G. W.: A General Theory of Elastic Stability, Wiley, London, 1973.
73. SUZUKI, Y. and OKUMURA, T.: Influence of cross-sectional distortion of flexural-torsional buckling, Final Report IABSE, New York, 1968.
74. KOLLBRUNNER, C. F. and HAJDIN, N.: Die Verschiebungsmethode in der Theorie der dunnwandigen Stabe und ein neues Berechnungsmodell des Stabes mit seinen ebenendeformierbaren Querschnitten. Publications IABSE, 28(II) (1968) p. 87.
75. FISCHER, M.: Das Kipp-Problem querbelasteter exzentrisch durch Normalkraft beanspruchter I-Trager bei Verzicht auf die Voraussetzung der Querschnittstreue, Der Stahlbau, 36 (1967) p. 77.
76. SCHMIED, R.: Die Gesamtstabilitat von zweiachsig aussermitting gedruckten dunnwandigen I-Staben unter Berucksichtigung der Querschnittsverformung nach der nichtlineare Plattentheorie, Der Stahlbau, 36 (1967) p. l.
77. RAJASEKARAN, S. and MURRAY, D. V.: Coupled local buckling in wide-flange beam-columns, J. *Struct. Div. ASCE,* 99(ST6) (1973) p. 1003.
78. JOHNSON, C. P. and VILL, K. M.: Beam buckling by finite element procedure, J. Struct. Div. ASCE,100(ST3) (1974) p. 669.
79. HANCOCK, G. J.: Local, distortional and lateral buckling of I-beams. J. Struct. Div. ASCE,104(STII) (1978) p. 1787.

80. HANCOCK, G. J., BRADFORD, M. A. and TRAHAIR, N. S.: Web distortional and flexural-torsional buckling, J. Struct. Div. ASCE,106(ST7) (1980) p. 1557.
81. IVÁNYI, M.: Interaction of stability and strength phenomena in the load carrying capacity of steel structures. Role of plate buckling, Doctor Tech. Sci. Thesis, Hungarian Academy of Sciences, Budapest, 1983 (in Hungarian).
82. HEGEDŰS, L. and IVÁNYI, M.: Interaction between plate and lateral-torsional buckling with regard to residual stresses, Periodica Polytechnica (Budapest), Civ. Eng., 29(3-4) (1985).
83. STOWELL, E. Z.: A unified theory of plastic buckling of columns and plates. NACA Technical Note No. 1556, 1948.

Chapter Four
84. MASSONNET, C. and MAQUOI, R.: Recent progress in the field of structural stability of steel structures, IABSE Periodica, 2/1978 (1978) 1.
85. GVOZDEV, A. A.: Ultimate Load Analysis of Structures, Gosstroizdat, Moscow, 1949. (In Russian.)
86. NAVIER, L. H.: Resume des Leçons Données à l'Ecole des Ponts-et-Chaussees, Paris, 1826.
87. KAZINCZY, G.: Experiments in fixed-end beams. Betonszemle, 2(1914) p. 68. (In Hungarian.)
88. BOLOTIN, V. V.: Reliability Theory and Structural Stability, University of Waterloo, Study No. 6,1952, p.385.
89. HALÁSZ, O. and IVÁNYI, M.: Tests with simple elastic-plastic frames, Periodica Polytechnica (Budapest),Civ. Eng., 23(1979) p. 157.
90. IVÁNYI, M., KÁLLÓ, M. and TOMKA,P.: Experimental investigation of full-scale industrial building section.Second Regional Colloquium on Stability of Steel Structures, Hungary, Final Report, 1986, pp.163-170.
91. IVÁNYI, M.: Interaction of stability and strength phenomena in the load carrying capacity of steel structures. Role of plate buckling, Doctor Techn. Sci. Thesis, Hungarian Academy of Sciences, Budapest,1983 (In Hungarian).
92. CLIMENHAGA, J. J. and JOHNSON, P.: Moment–rotation curves for locally buckling beams, J. Struct. Div. ASCE, 98(1972) ST6.
93. IVÁNYI, M.: Yield mechanism curves for local buckling of axially compressed members, Periodica Polytechnica (Budapest),Civ. Eng., 23(3-4) (1979) 203-216.
94. IVÁNYI, M.: Moment–rotation characteristics of locally buckling beams, Periodica Polytechnica (Budapest), Civ. Eng., 23(3-4) (1979) 217-230.
95. GIONCU, V., MATEESCU, G. and ORASTEANU, S.: Theoretical and experimental research regarding the ductility of welded I-sections subjected to bending, in: Stability of Metal Structures, Proceedings of the 4th International Colloquium on Structural Stability, Asian Session, Beijing, 1989, 289-98.
96. HORNE, W. R.: Instability and the plastic theory of structures, Trans. EIC, 4 (1960) p. 31.
97. THURLIMANN, B.: New aspects concerning the inelastic instability of steel structures, J. Struct. Div. ASCE, 86 (1960), ST1, 99-120.
98. HALÁSZ, O.: Limit design of steel structures. Second-order problems. D.Sc. Thesis, Budapest, 1976 (In Hungarian).

99. DRUCKER, D. C.: A more fundamental approach to plastic stress–strain relations, in: Proc. 1st US National Congress of Applied Mechanics, ASME, 1951, p. 487.
100. DRUCKER, D. C.: On the postulate of stability of material in the mechanics of continua, J. Mech. (Paris), 3 (1964), p.235.
101. MAIER, G.: Sull'equilibrio elastoplastico delle strutture reticolari in presenza di diagrammi forze-elongazioni a trotti desrescenti, Rendiconti, Instituto Lombardo di Scienze e Letture, Casse di Scienze A, Milano, 95 (1961) p. 177.
102. MAIER, G. and DRUCKER, D. C.: Elastic–plastic continua containing unstable elements obeying normality and convexity relations, Schweiz. Bauzeit, 84(23) (1966).
103. IVÁNYI, M.: Effect of plate buckling on the plastic load carrying capacity of frames. Paper presented at IABSE 11 Congress, Vienna, 1980.
104. BAKSAI, R.: Plastic analysis by theoretical methods of the state change of steel frameworks. Diploma Work, Techn. Univ. Budapest, 1983 (in Hungarian).
105. SZABÓ, J. and ROLLER, B.: Theory and Analysis of Bar Systems, Műszaki Könyvkiadó, Budapest, 1971 (in Hungarian).
106. HORNE, M. R. and MERCHANT, W. The stability of frames. Pergamon Press, 1965.

FATIGUE OF STEEL PLATED STRUCTURES

K. Yamada
Nagoya University, Nagoya, Japan

Abstract

When steel plated structures are subjected to repeated stress cycles, they are to be designed for fatigue. Various fatigue tests with small or large scale test specimens are carried out, some of which are presented in the text as examples. Fatigue assessment procedures based on design S-N curves and fracture mechanics approach are demonstrated with examples. Finally, fatigue life evaluations of existing steel plated structures based on service stress measurements are described.

1. Introduction

1.1 Fatigue of Steel Structures

Steel plated structures are normally designed for three limit states, namely;
(a) ultimate limit state,
(b) serviceability limit state,
(c) fatigue limit state.
In order to assure the long term safety of the steel structure, it must be properly designed for one or more extreme load scenarios which are expected during the design life.

Durability of the steel structure is also an important design factor to be considered. Corrosion and fatigue are the two main factors which determine the durability of the steel plated structures. Fig. 1.1 shows an example of excessive corrosion, which occured on the lower flange of a cross beam of a highway truss bridge. Debris coming from the concrete deck kept this place humid, and continuous corrosion took place. Negligence of maintenance during and after World War II also contributed to corrosion. Careful detailing and periodic repainting may reduce such corrosion problems.

When steel structures are subjected to repeated loading, and when the stresses and their number of cycles exceed certain limits, fatigue cracks initiate and propagate from points of which stress concentration [1,2,3]. This causes severe maintenance problems for many steel plated structures.

If the fatigue limit state governs design, it may also lead to failure of the structure. Designers must avoid fatigue problems by understanding both the fatigue behavior of steel plated structures, and the fatigue design procedure.

Various kinds of fatigue cracking have been experienced for different types of steel plated structures, such as highway and railroad bridges [1], offshore structures [2], and crane runway girders [3]. As an example, Fig. 1.2 shows a fatigue crack emanating from the edge of a stringer supporting the concrete deck of a highway bridge. The excessive stresses which caused the fatigue crack seem to have occured two reasons. Firstly, deterioration of the concrete slab may have reduced load distribution, so that wheel loads were directly transferred to this stringer. Secondly, overloaded trucks were frequently observed on this highway, which connects two industrial centers, and these caused high stresses in the stringer.

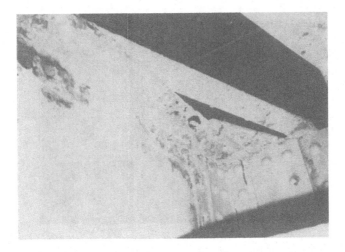

Fig.1.1 Excessive corrosion of lower flange of cross beam.

Fig.1.2 Fatigue crack emanating from scallop of stringer.

Another example is shown in Fig. 1.3, where a large crack was found in a groove welded box section of a crane girder located in a steel works. For an unknown reason, a partial penetration groove weld was specified during design of the box section. This detail is susceptible to fatigue cracks, because the unwelded part of the groove weld may act as crack initiator. Such a detail is therefore not normally used when a structure is expected to be subjected to a large number of stresse cycles.

(a) Fatigue cracked box section of crane girder.

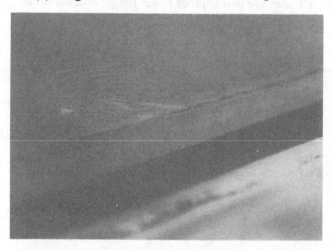

(b) Close-up view of fatigue cracked lower flange.

Fig.1.3 Fatigue crack occurred in box section of crane girder.

The main factors which govern fatigue of steel plated structures are;
(a) structural details,
(b) stress ranges,
(c) number of cycles.
Therefore, in order to avoid fatigue cracks in steel plated structures, favorable structural details should be used, if possible, and acurate estimates must be made for stress ranges and number of cycles.

Numerous research papers related to fatigue of metallic materials and steel plated structures can be found in the literature. However, only related references are listed herein, and readers can refer to appropriate textbooks, some of which are listed in the references, if necessary.

1.2 Basis of Fatigue Design

(1) Stress Cycles

Steel plated structures may be subjected to various kinds of loads, which lead to stress cycles. Some stress cycles are shown schematically in Fig.1.4. When the magnitude of the loads is constant, constant amplitude (CA) stress cycles are observed, as shown in a). The stress ratio $R=S_{min}/S_{max}$ depends on whether the minimum stress S_{min} is high or low. If S_{min} is negative, as shown in b), this gives a negative stress ratio. A value of R=-1 indicate that a complete stress reversal takes place, such as occured when a round bar specimen is rotated during a bending test.

When a bridge is subjected to truck loads, stress cycles such as those shown in c)-e) may be observed. Stress ranges vary according to weight of trucks. These are known as variable amplitude (VA) stress cycles. Normally, stress ranges due to vehicles (live load) are added to the minimum stress, which is the stress due to dead load, such as shown in c). Sometimes, residual stresses due to welding modify the minimum stress and may lead to stress cycles such as shown in d). The stress cycles shown in e) are of variable amplitude, so that both stress range and stress ratio changes. When some stresses become negative, this can be expressed schematically as shown in f).

Steel plated structures are subjected to different stress cycles according to the type of structures and its structural members. A good example is a truss bridge on the Shinkansen line, where bullet trains of 16 carriages have been running since 1964. Each carriage is about 25 m long. When the train, of total length approximately 400 m, passes over the long span truss bridge, with a span of for example 60 m, the upper and lower chords are subjected to one large stress cycle with small ripples on top of it. However, stringers supporting rail sleepers have a short span, for example 6 m, and the passage of each carriage causes a stress cycle. This results in 17 cycles per passage of a train. Assuming that 120 trains pass daily on this bridge, the resulting number of stress cycles is as follows.

The upper or lower chords :

120 (trains) × 365 (days) × 70 (years)   = 3,066,000 cycles

The stringers:

17 × 120   ×365   ×70   = 52,122,000 cycles

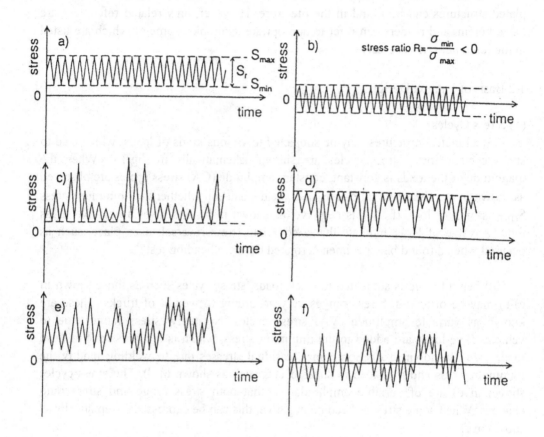

Fig.1.4 Various stress cycles observed in steel plated structures.

(2) Fatigue Design Recommendations

Fatigue design can be carried out using the fatigue design codes, such as,

a) BS5400, Part 10   (1980)
b) ECCS Fatigue Design Recommendations   (1985)
c) Eurocode 3, Chapter 9, Fatigue   (1993)
d) JSSC Fatigue Design Recommendations   (1993)

Typical design S-N curves are shown in Fig.1.5. Since all these fatigue design codes are based on recent fatigue research and test data, they have several features in common;

(a) A series of S-N curves are given, and structural details are categorized in groups according to their fatigue behavior (stress category).

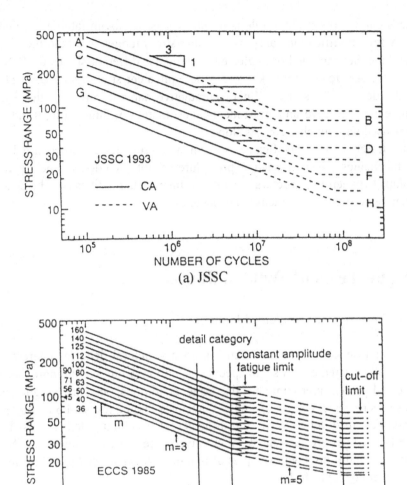

Fig. 1.5 Typical design S-N curves.

(b) Slopes of S-N curves are basically taken as k=3, with minor modifications down to a constant amplitude (CA) fatigue limit.

(c) The rain flow counting method (or reservoir method) is normally used to count the number of cycles and the stress ranges for variable amplitude (VA) loading.

(d) Miner's cumulative damage rule is normally used to obtain fatigue life under VA loading. A few modifications are made so that stress ranges less than the VA cutoff limit may be either included or neglected, or they use different slopes for S-N curves with VA stresses, for example k'= k+2. The equivalent stress range concept is also used in the design codes. The Miner's cumulative damage rule, and the equivalent stress range concept, give an almost identical estimation of fatigue life, if the same S-N curve is used without changing its slope.

(e) Some design codes specifically mention a fatigue design procedure based on fracture mechanics. Even if such a procedure is not included in the design code, relationships between the structural details and their weld qualities and the design S-N curves can be verified using fracture mechanics.

2. Fatigue Tests of Welded Joints

2.1 Fatigue Tests of Various Welded Joints

In order to establish new S-N curves for a structural detail of interest, fatigue tests are usually carried out. A set of specimens, normally numbering more than seven , are fabricated under approximately the same conditions, and are fatigue-tested at different stress range levels. An example of a tensile fatigue test set-up is shown in Fig. 2.1. Sometimes, in order to verify the influence of a certain detail, several kinds of fatigue test specimen are fabricated. Alternatively, the same specimens can be tested under various conditions. Two examples of fatigue tests on welded joints carried out at Nagoya University are as follows:

(1) Tensile Plate with Side Gussets, as-welded :
The first example considers fatigue tests on four different types of specimen [5,6]. In order to see the influence of attachment length (L) on the fatigue strength of tensile plates with side gussets, gussets with L= 50, 100, and 200 mm groove-welded on both sides of 200 mm wide, 10 mm thick tensile plates, as shown in Fig. 2.2. Specimens were designated G5, G1 and G2. An additional set of specimens with 200 mm long gussets transversely fillet welded on the sides of the plates was also tested. These were designated T2 specimens.

Fatigue cracks initiated and propagated from the toes of the longitudinal welds at the end of the side gussets, where stress concentrations were high. Stress concentrations, and therefore the fatigue strength, are influence by the attachment

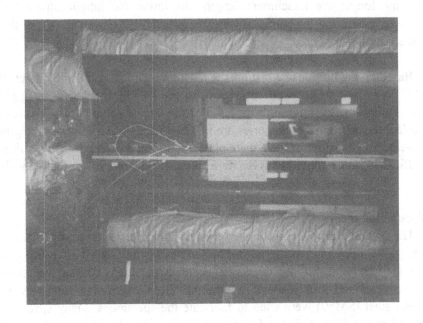

Fig.2.1 Fatigue test of tensile specimen with longitudinal
gussets fillet welded to center of plate.

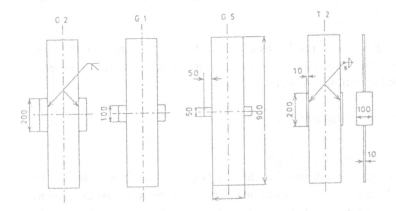

Fig.2.2 Tensile specimens with side gussets of various sizes.

length. Fatigue test results shown in Fig. 2.3 clearly show the effect of the attachment length; the longer the attachment length, the lower the fatigue strength. The T2 Specimens, where the attachments were transversely fillet-welded, show the lowest fatigue strength.

Based on the fatigue test results, the tensile specimens with side gussets (G5, G1 and G2) may be categorized according to the ECCS and JSSC fatigue design recommendations, as shown in Fig. 2.4. For the tensile specimens with side gussets, as-welded, the appropriate categories are ECCS-45 and JSSC-H. When the gussets are fillet welded on the sides of the tensile plate, their fatigue strength is lower than ECCS 45 or JSSC-H, so design S-N curves for a category one rank lower than that should be used.

(2) Weathering Steel :
The second example considers a series of fatigue tests on non-load carrying fillet welded joints following long term atmospheric exposure [7,8]. Tests were carried out on non-load carrying, 80 mm wide fillet welded specimens without weathering, and after two years and four years of weathering. Both weathering steel (SMA50) and structural steel (SM50) were used to fabricate the specimens. Many specimens were fabricated at the same time, and they can be considered as having the same initial conditions. The specimens were then exposed to atmospheric conditions on racks facing south with a slope of 30°, for a pre-defined period (see Fig.2.5). The weathering was rather favorable to the unpainted weathering steel, because occasional rain washed away dust from the steel surface, and the sunshine kept the surface relatively dry.

The fatigue test results, as shown in Fig.2.6, indicate that the fatigue life of each specimen after atmospheric exposure was about the same or longer than the non-weathered control specimens.

2.2 Allowable Stress Ranges

(1) Constant Amplitude Fatigue
In order to establish allowable S-N curves for fatigue design, test data for the structural detail of interest must be collected. Results of past experiments can be used, or fatigue tests of the detail have to be carried out. Normally the test data is plotted on an S-N curve. If a set of test data is available, then linear regression analysis is normally carried out to define a mean S-N curve, standard deviation (s), and correlation coefficient (r), as shown for example in Fig.2.6.

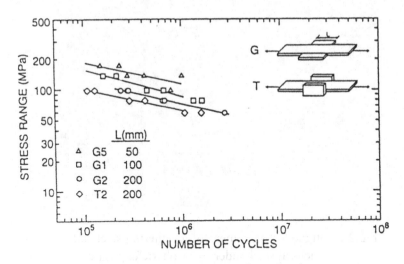

Fig.2.3 Fatigue test results of tensile plate
with various side attachments.

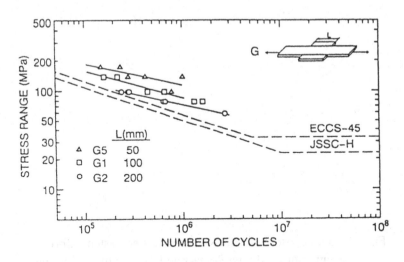

Fig.2.4 Comparison with the allowable stress ranges.

Fig.2.5 Fatigue test specimens of weathering steel and
structural steel under atmospheric exposure.

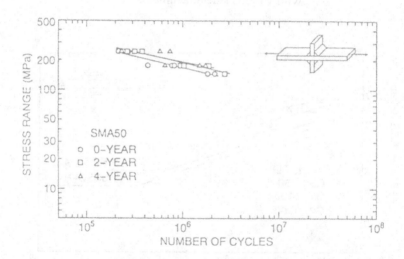

Fig 2.6 Fatigue test results for weathering steel tested before
weathering, and after 2-years and 4-years of weathering.

$$logN_f = A - BlogS_r \tag{2.1}$$

where N_f is the number of cycle to failure, S_r is the applied nominal stress range, and
A and B are the regression coefficients.

Allowable S-N curves are normally determined by shifting the mean S-N curves two standard deviations away from the mean. This corresponds to approximately A 97.7 % survival of the test data.

For a fatigue design code, a great effort is normally made to collect and evaluate large numbers of test data generated in the past, for example in the form of a fatigue test data base. An example of a fatigue test data base system developed for personal computers is described in Ref.9. Having established the data base, the test data is compared with a series of pre-defined design S-N curves, and the stress category for each structural detail is determined. A design S-N curve is selected so that no test data indicates less fatigue strength than the design curve. Fig.1.5 shows design S-N curves as specified in JSSC and ECCS Fatigue Design Recommendations [11,12]. The JSSC Fatigue Design Specification specifies eight parallel, equally spaced curves, for weldments subjected to normal stress. A jump of one curve, for example category C to D corresponds to approximately doubling of the fatigue life. The ECCS Fatigue Design Recommendations use twenty equally spaced S-N curves. In this case a jump between curves corresponds to approximately a 50% increase in the fatigue life.

Design S-N curves are based on constant amplitude (CA) fatigue test data. Only limited test data is available in the long life region, for example more than ten million cycles, so determination of the CA fatigue limits is a rather difficult task, if not impossible. The CA fatigue limits are determined as a lower bound of a few available long life test results. In order to simplify the design recommendation, the CA fatigue limit is set where the design S-N curves attain 10^7 cycles [10] or 5×10^6 cycles [11]. Fracture mechanics analyses of fatigue crack propagation life are also helpful in determing the CA fatigue limit. This will be discussed in Chapter 4.

(2) Variable Amplitude Fatigue

When the structure is subjected to variable amplitude (VA) stress cycles, the design S-N curves are more complicated. The basic concept of the design procedure for VA fatigue is to use the well known Miner's rule, or the equivalent stress range concept, which are based on the linear cumulative damage rule.

Miner's rule can be expressed by the following equation.

$$\sum_{i=1}^{n} \frac{n_i}{N_i} = 1.0 \tag{2.2}$$

where n_i is the number of cycles of the ith stress range S_i, and N_i is the fatigue life due to S_i ($i=1,2,\cdots,n$), which is obtained from an S-N curve for the detail of interest.

A so-called equivalent stress range ($S_{r.eq}$) can be obtained by modifying Eq.2.2. This gives the same damage as the application of a stress range S_i over a number of cycles n_i, when the number of cycles of $S_{r.eq}$ is equal to .

$$S_{r.eq} = \left(\sum_{i=1}^{n} \frac{n_i S_i^k}{N} \right)^{1/k}$$ (2.3)

where k is the inverse slope of the S-N curve. Usually k=3 is used, and the equivalent stress range becomes a so-called "root-mean-cube" stress range, as follows.

$$S_{r.eq} = \left(\sum_{i=1}^{n} \frac{n_i S_i^3}{N} \right)^{1/3}$$ (2.4)

Linear cumulative damage rules, such as expressed by Miner's rule (Eq.2.2) or the quivalent stress range (Eq.2.3), are valid only when the linear part of the S-N curve on a log-log scale is considered. When the number of stress cycles less than the CA fatigue limit are large in the stress range histogram of linear cumulative damage rules often lead to an over conservative estimate of fatigue life.

Design S-N curves, especially in the long life region, sometimes vary according to the design code. Difference are mainly due to the effect of the large number of lower stress ranges in the VA stress cycles. In the present fatigue design recommendations the following S-N curves are used for VA fatigue design.
a) The CA S-N curves are linearly extended for VA S-N curves. The design concept is often called the Modified Miner's rule.
b) The CA S-N curves are linearly extended to the VA cut-off limit, below which the effect of stress cycles is neglected.
c) The slope of S-N curve is modified at the CA fatigue limit, from k to k=k'+2, in order to define VA design S-N curve in the longer life region.
d) The VA cutoff limit is also introduced into case c), when the effect of a large number of small stress ranges below the VA cut off limit is neglected.

The accuracy of these procedures has yet to be clarified. Only limited VA fatigue test data in the long life region is available, and the large scatter of results makes it difficult to say which procedure most accurately expresses the test data more' accurately.

2.3 Summary

Two examples of fatigue tests on small scale tensile specimens are described in this paper. These are a) 10 mm thick, 200 mm wide tensile plates with welded attachments of various sizes, and b) non-load carrying 80 mm wide fillet welded specimens of weathering steel and structural steel, tested before weathering and after weathering for two years and four years. Data is compared with allowable stress ranges specified in ECCS and JSSC Fatigue Design Recommendations.

The procedure used to define allowable stress ranges for fatigue design is also explained. Fatigue design procedures described in various fatigue design recommendations are summarized. Some steel plated structures, such as highway bridges subjected to heavy truck traffic or railway bridges on busy lines, are often subjected to a very large number of stress cycles at low stress range levels. There are some differences in fatigue design procedures used to estimate fatigue life for such a long life loading. Further research is needed to carry out fatigue tests of weldments under variable amplitude stress in the long life region. Approximate analytical procedures need to be developed to back-up the test data, and one promising procedure is to use the fracture mechanics approach.

3. Fatigue Tests of Welded Structures

3.1 Full Scale Fatigue Tests

Although the majority of data generated in the past was for small scale specimens, mainly in tension, large scale fatigue testing machines with a large loading capacity have made large scale fatigue tests possible. Fig.3.1 shows one of the largest fatigue testing machines with a load capacity of 400 tonf. It was specially developed to carry out fatigue tests on full scale specimens for the Honshu-Shikoku Bridge project, one of the largest bridge construction projects in the world during this century. In 1988 a series of suspension bridges and cable stayed bridges connected two main islands of Japan, the Honshu and the Shikoku, and the Akashi Kaikyo bridge, with a span of 1990 m, is now under construction.

The Honshu-Shikoku Bridge Authority (HSBA) has used the fatigue testing machine continuously since the 1970s for various fatigue test programs, some of which are as follows.
a) Fatigue tests of large longitudinal butt welds with partial penetration [13].

(a) Full-scale truss member

(b)Full scale section of cable stayed bridge

Fig.3.1 Fatigue tests of full-scale specimens carried out
by the Honshu-Shikoku Bridge Authority.

b) Fatigue tests of a truss made of high strength steel [14].

c) Effect of stress ratios on the fatigue strength of cruciform fillet welded joints [15].

d) Fatigue tests of a full-size truss chord member [16].

e) Fatigue tests of large gusset joints using 800 MPa class steels [17].

Test results were used in fatigue design of the long span bridge made of high strength steel.

The advantages of using large-scale fatigue test specimens are as follows:

a) Since the specimens are designed and fabricated in a similar way to the actual structures, the fatigue cracks and the fatigue life obtained from the full scale test specimens can be directly used for the design of the actual structures.

b) Even though the specimens have to be modified due to the limitations associated with the testing conditions, the specimens seem to reflect the actual structures better than small-scale coupon tests.

c) It seems that the full scale test specimens reflect the actual structures in terms of plate thickness, welding residual stresses, welding technique, the effect of 3-D plate connections, and/or the fabrication process. It is not always possible to introduce these details into small-scale test specimens.

d) It may be possible to introduce various joint details in a large scale specimen, so that numerous fatigue test data can be obtained from one specimen. In this case fatigue cracks which initiate in a structural detail have to be stiffened or repaired to stop further fatigue crack propagation. Detection of fatigue cracks is therefore very important for full-scale fatigue tests of specimens incorporating various structural details.

3.2 Fatigue Tests of Large Size Specimens

The full-scale fatigue tests carried out by the Honshu-Shikoku Bridge Authority were very valuable, and gave a great deal of information for designers to make detail drawings of their bridges. They were, however, very costly, and it was almost impossible to carry out several tests on the same specimens. Full-scale fatigue tests are not always possible for other research institutes because of the limitations of their fatigue testing machines. Even so, fatigue tests on large specimens can be carried out with fatigue testing machines of smaller capacity.

(1) Gussets Welded to Tension Flange

An example of the large-scale fatigue tests carried out at Nagoya University is shown in Fig. 3.2 [19,20,21]. A 4 m long beam was subjected to downward loading at mid-span using 35 tonf actuator. Various attachments were welded in the tension flange and the web, as shown in Fig. 3.3. The fatigue test of the beam was carried out with a loading speed of about 1 Hz, which is much slower than the tensile fatigue tests

Fig.3.2 Fatigue test of a large size beam.

Fig.3.3 Various attachments welded to the tension flange.

of plates with side gussets. Large mid-span deflections made the fatigue test slower.

Fatigue cracks were found at the toe of the fillet weld at the end of the gussets welded to the tension flange. Because the beam was subjected to a concentrated load at mid-span, the stresses in the tension flange varied with position in the span. Fatigue cracks therefore developed first near mid-span. These cracks were retrofitted by drilling a stop-hole at the end of the cracks, and adding splice plates with friction-grip high strength bolts when necessary. Fatigue tests could then be continued. By repeatedly doing this, the beam produced about twenty fatigue cracks, that were observed initially in highly critical sections, and then gradually in less stressed sections.

Fatigue test results for 200 mm long side gussets welded to the tension flange of a beam are plotted in Fig.3.4 [20]. Also plotted are fatigue test data for 200 mm wide tensile plates with as-welded 200 mm long gussets. The mean S-N curve and the standard deviations (s) were computed from a total of 26 results, as follows:

$$logN_f = 12.699 - 3.648logS,$$
$$s = 0.123$$
(3.1)

The mean S-N curve and the confidence limits plotted two standard deviations away from the mean are shown by solid lines. The fatigue strength of this type of gusset is known to be low, and the ends of the gussets are normally specified to be ground to a radius of r≥40 mm for railway bridges in Japan in order to avoid the possibility of fatigue cracks. As-welded gussets are often used for steel structures other than railway bridges and the JSSC recommendation for fatigue design specifies this detail as class H. All test data seems to satisfy the allowable stress range, as shown in Fig.3.4

(2) Groove Weld Repair of Crack

After the fatigue tests in the as-welded condition, the test beams had a number of fatigue cracks in the tension flange. These fatigue cracks were about 20 mm long and stop-holes were drilled at the crack tips. The crack were repaired in the following way. As shown in Fig.3.5, the original cracks were lengthened to 100 mm by drilling and gas cutting. Grooves were then made on both sides of the plates by arc air gouging in the flat position. Plate were groove-welded with four or five passes. A 3 mm thick and 8 mm diameter penny-shaped disc was tack-welded at the drilled holes in order to avoid the melting down of the weld metal. The groove weld was then back-chipped from underneath. A dye check was used to ensure that no crack was left at the weld, before the groove weld was made in the overhead position. The final repair weld was made in the overhead position to simulate unfavorable site conditions, where the weld repair must be made from underneath. This welding position is believed to cause more weld defects than the flat position. The weld reinforcement was then ground flush, and the end weld of the gusset was ground to a radius of about 20 mm.

(3) Fatigue Test

The fatigue tests after applying repair welds were carried out in the same manner as for the as-welded specimens. The beams were monitored periodically and the

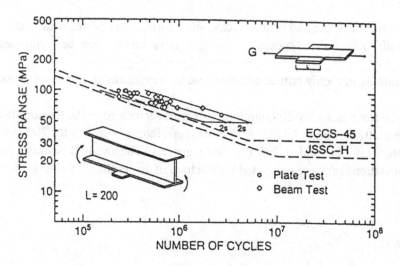

Fig.3.4 Fatigue test results for 200 mm long gussets welded to
tension flanges of beams and tensile specimens.

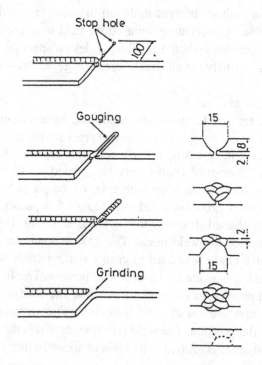

Fig.3.5 Repair of crack at gusset end by groove welding.

fatigue life (N_f) was defined when a crack propagated to a length of approximately 10 mm. When fatigue cracks became about 20 mm long, a stop-hole was drilled and stiffened by high strength bolted splices.

Eight out of 20 repair welds exhibited fatigue cracks. These cracks initiated from the fillet weld of the gusset ends and propagated in the direction perpendicular to the principal stress. Typical fatigue fracture surfaces are shown in Fig.3.6. Fig.3.6A shows a fracture surface with few weld defects. The fatigue crack seems to originate from grinding marks at the fillet, and it showed the longest fatigue life among the tested specimens. In Fig.3.6B, blowholes exist near the fillet, and the fatigue crack initiated from these blowholes. This resulted in an average fatigue life compared with other results. Most harmful to fatigue resistance were large and numerous defects near the fillet, as shown in Fig.3.6C. Fatigue cracks seem to originate from these defects, and to have coalesced to a large crack. This specimen showed the shortest fatigue life.

Fatigue test results of G-type gussets after weld repair was made are plotted in Fig.3.7. All test data indicate a longer fatigue life than the as-welded results shown by a confidence interval. This is because the ends of the gussets were ground to a radius of approximately 20 mm, which eliminated high stress concentrations at the weld toes. Results imply that the weld defects observed in the repair welds were less harmful than the stress concentrations at the weld toes of the gusset ends in the as-welded condition. Therefore, the repair welds described here are acceptable for repairing fatigue cracks and to prolong the fatigue life, assuming the required life of the structure is properly known by the designer.

3.3 Summary

A full scale fatigue tests, such as those carried out by the Honshu-Shikoku Bridge Authority, are extremely valuable, because they include variables which cannot always be incorporated in small-scale test specimens. However, a full-scale tests are usually very costly, and require testing machines with a large loading capacity. Fairly large-scale fatigue tests, such as bending tests of beams, are easier to be performed using fatigue testing machines of moderate capacity. These tests can also simulate actual conditions of steel plated structures. By including various welded details in one specimen it is possible to obtain multiple cracks, provided that fatigue cracks are detected at an early stage and they are properly retrofitted to prevent further crack propagation.

Large-scale fatigue tests often show somewhat different fatigue strength when compared with small-scale specimens of the same kind. This is due to the effects of residual stresses and/or the size effect, which are normally known as plate thickness

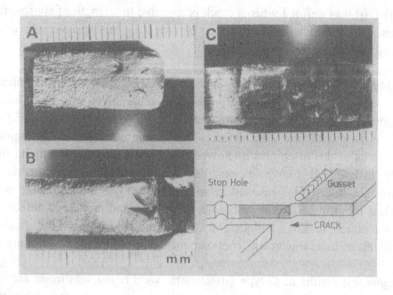

Fig.3.6 Typical fatigue cracked fracture surfaces of weld repaired gusset ends.

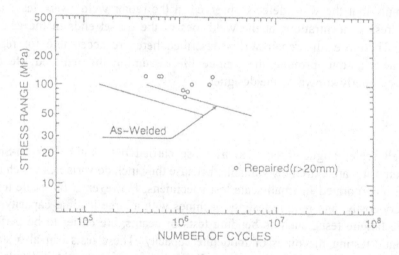

Fig.3.7 Fatigue test results of beams with 200 mm
long gussets, after weld repair.

effect. Careful examination of test data from full-scale or large-scale specimens will enable the data from a large number of small-scale test specimens to be correlated to actual steel plated structures.

4. Fracture Mechanics Approach to Fatigue

4.1 Fundamentals of Fracture Mechanics

Since fatigue involves the processes of crack initiation and propagation, a successful effort was made to correlate it to the characteristics of the leading edge of the crack. Linear elastic fracture mechanics defines quantitatively the severity of a crack by the stress intensity factor K as proposed by G.R.Irwin. It has been widely applied to solve various engineering problems related to cracks, such as brittle fracture, fatigue, stress corrosion cracking [1,23].

One of the most common applications deals with fatigue crack growth. Since P. Paris proposed an empirical equation to express the fatigue crack growth rate (da/dN) as a function of the stress intensity factor range (Δ K), it has been widely used in analysis of fatigue crack propagation life, fracture analyses of actual failures, design specifications, etc. The so-called "Paris equation" is expressed as follows.

$$da \quad dN = C (\Delta K)^m \tag{4.1}$$

where C and m are the material constants. They are determined experimentally from cracked specimens, for which Δ K is known. Eq.4.1 is linear on a log-log scale, as shown in Fig.4.1.

The stress intensity factor range (Δ K) is conveniently expressed by Eq.4.2.

$$\Delta K = F(a) \cdot S_r \sqrt{\pi a} \tag{4.2}$$

where a is the crack size, and F(a) is a correction factor to account for the difference between the crack of interest and the so-called "Griffith crack".

Substituting Eq.4.2 into Eq.4.1, the following relationship can be obtained. This relationship allows fatigue crack propagation life (Np), for propagation from an initial crack size of a_i to a final crack size of a_j, to be calculated.

$$N_p = \int_{a_i}^{a_j} \frac{1}{C \{ S_r \sqrt{\pi a} \cdot F(a) \}^m} \, da \tag{4.3}$$

It is well known that when the applied stress range S_r is constant, Eq.4.3 can be written as follows.

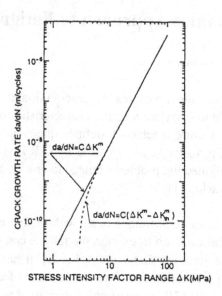

Fig.4.1 Fatigue crack growth rates of structural steel.

$$N_p = \left(\int_{a_i}^{a_f} \frac{1}{C\{\sqrt{\pi a} \cdot F(a)\}^m} \, da \right) \cdot S_r^{-m}$$

(4.4a)

$$= A \cdot S_r^{-m}$$

(4.4b)

Taking the logarithm of Eq.4.4, the following equation can be obtained:

$$logN_p = logA - m \cdot logS_r$$

(4.4c)

Eq.4.4c shows the same form as an S-N curve. For structural steel, both the material constant m, and the slope of the S-N curve (B) for welded joints are approximately equal to 3. Small fatigue cracks actually initiate at an early stage of fatigue in the welded joints, where crack-like defects or high stress concentrations exist. These reduce the initiation stage of fatigue and make it relatively less important than the propagation stage. Therefore, Np seems to predict a large part of the total fatigue life of welded joints.

4.2 Fatigue Crack Growth Rates

Numerous fatigue crack growth rate measurements have been carried out since P. Paris proposed Eq.4.1. A large scatter of fatigue crack growth rates was usually observed in the early measurements, owning to differences in shapes and sizes of specimens, the lack of an accurate stress intensity factor expression, unawareness of crack closure effects, etc.

The Paris equation was then modified based on the following considerations and more accurate experimental results. The main facts considered were as follows.
(1) The fatigue crack growth rate gradually deviates from Eq.4.1, and finally becomes zero, when ΔK decreases. The fatigue crack does not propagate when ΔK is less than the threshold value of the stress intensity factor range, ΔK_{th}.
(2) Fatigue crack growth rates and ΔK_{th} are largely influenced by the stress ratio and residual stresses, because the crack closure influences growth rates. The crack closure occurs when plastic elongation, which develops near the crack tip, remains at the crack surface after crack propagation, and closes the crack during the unloading stage. This behavior is called crack closure. It reduces the effectiveness of the applied stress in terms of crack growth.

A more general expression for fatigue crack growth rates is as follows.

$$da\ dN = C\left(\Delta K^m - \Delta K_{th}^m\right) \quad when\ \Delta K \geq \Delta K_{th} \tag{4.5a}$$

$$da\ dN = 0 \quad when\ \Delta K \leq \Delta K_{th} \tag{4.5b}$$

Intensive research was carried out by the National Research Institute for Metals of Japan(NRIM) to obtain crack growth rates for various structural steels. Different base metals and attachment using various welding procedures were considered. The results were published as a data base [24]. Based on experimental results, JSSC Fatigue Design Recommendations proposed the following material constants for structures steel as follows[12]:

$$C = 1.5 \times 10^{-11} \tag{4.6a}$$

$$m = 2.75 \tag{4.6b}$$

$$\Delta K_{th} = 2.9 \tag{4.6c}$$

4.3 Stress Intensity Factor Expression in Engineering Problems

A value of ΔK must be caluculated in order to use linear elastic fracture mechanics for the fatigue crack propagation analysis. Although numerous ΔK expressions for various crack problems have been derived and can be found in handbooks [25], Δ K expressions for welded details, which are essential for plated steel structures, are not always available. The finite element technique using a singular crack tip element is available, but requires a laborious preparation of input data. Moreover, it may not be easy to use when crack size and shape change, as is the case for the fatigue of weldments. The objective of this section is therefore to present a rapid method of calculating K values for the opening mode of cracks in structural details [26]. This technique has been successfully used for various applications.

(1) Some Useful Solutions for Correction Factors

It is convenient to write the correction factor, $F(a)$, for the Griffith crack in terms of a combination of various correction factors[26].

$$F(a) = F_s \cdot F_E \cdot F_T \cdot F_G \qquad (4.7)$$

where F_S accounts for the influence of the free surface, F_E for the influence of an elliptical crack front, F_W for the finite width or thickness, and F_G for the non-uniform opening stresses. Some known solutions can be used, for each individual factor. Their combined use to estimate K values should for various crack problems may involve some approximations, but any errors should be small.

For two-dimensional crack problems, such as shown in Fig.4.2, the following expressions are given:

$$F_s = 1.12 \qquad (4.8)$$

$$F_w = \sqrt{W \cdot \pi a \cdot tan(\pi a \cdot W)} \qquad (4.9)$$

The geometry correction factor, F_G , accounts the effect of non-uniform stresses at the crack surface, which will be discussed later.

For three-dimensional crack problems, such as shown in Fig.4.3, the following expression is given for K at the crack front which intersects the minor axis for the elliptical crack in an infinite body.

$$F_E = \frac{1}{E_k}, \quad E_k = \int_0^{\pi \cdot 2} \left[1 - \left(1 - a^2 \cdot b^2 \right) sin\,\theta \right]^{1 \cdot 2} d\theta \qquad (4.10)$$

where E_k is the complete elliptical integral of the second kind, and a and b are the length of the minor and major semi-axes of the elliptical crack respectively. For a part-through semi-elliptical surface crack, an expression which accounts for the restraint

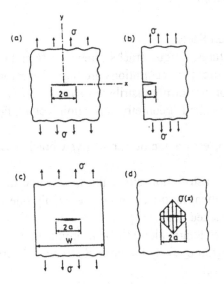

Fig.4.2 Two-dimensional crack models; (a) Griffith crack, (b) Edge crack;
(c) Central crack in finite width plate, (d) Central crack subjected
to non-uniform opening stress.

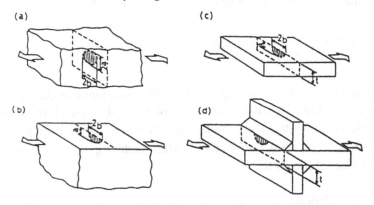

Fig.4.3 Three dimensional cracks; (a) Elliptical crack in infinite body,
(b)Semi-elliptical surface crack, (c) Semi-elliptical crack in
finite plate, (d) Semi-elliptical crack from weld toe.

effect of the net ligament around the part-through crack should preferably be used:

$$F_s = 1.12 - 0.12(a \cdot b) \qquad\qquad (4.11)$$

$$F_W = \sqrt{W \cdot \pi a \cdot \tan(\pi a / W)}$$ (4.12)

(2) Geometry Correction Factor, F_G

For welded structures, fatigue cracks often initiate and propagate from geometric discontinuities, where stress concentrations are high. Therefore, it is essential to allow for the influence of non-uniformly distributed stresses on K values. This influence is accounted for by the so-called "geometry correction factor", F_G.

The geometry correction factor for a given crack size can be determined as follows:

(1) Compute the stress distribution along the line where the crack shall be inserted. Any available finite element method can be used. Only one finite element analysis is carried out for the uncracked body.
(2) Insert a crack of given length along the appropriate line.
(3) Compute K by integrating away the normal stresses, determined in step 1, applied over the length of the crack.
(4) Repeat steps 2 and 3 for any desired crack size.

The value of K in step 3 can be computed from Eq.4.13.

$$K = \sqrt{\pi a} \, \frac{2}{\pi} \sum_{i=1}^{n} \sigma_{bi} \int_{b_i}^{b_{i+1}} \frac{1}{\sqrt{a^2 - b^2}} \, db$$ (4.13)

where the discrete stress σ_{bi} represents the stress distribution in the uncracked body, which can be computed by the finite element method. Eq.4.13 is derived from a known solution for a Griffth crack subjected to two pairs of splitting forces applied at the crack surface, as shown in Fig.4.4. After factoring out the mean stress, σ, integration of Eq.4.13 leads to:

$$K = \sigma\sqrt{\pi a} \cdot \frac{2}{\pi} \sum_{i=1}^{n} \frac{\sigma_{bi}}{\sigma} \left(arcsin\frac{b_{i+1}}{a} - arcsin\frac{b_i}{a} \right)$$ (4.14)

The second part of the right hand side of Eq.4.14 accounts for the influence of the non-uniformly distributed stresses due to geometrical stress concentrations, and is therefore defined as the geometry correction factor, F_G [6].

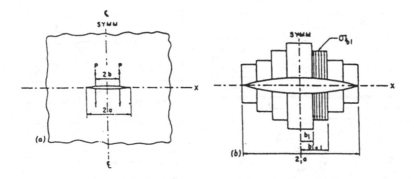

Fig.4.4 Computation of F_G. Central crack subjected to, (a) two-pairs of splitting forces; (b) non-uniform stresses.

4.4 Fatigue Crack Propagation Analysis

Fatigue crack propagation analyses have been carried out for various welded details, such as non-load carrying fillet welded specimens [27,28,29], tensile plates or flange with in-plane side gussets [30,31], tensile plates with out-of-plane gussets of weathering steel [28,32], various welded details of sandwich steel plates [33], and longitudinal fillet welded specimens. In order to demonstrate an example of the fatigue behavior of welded joints, and the analysis of fatigue crack propagation life using fracture mechanics, fatigue analyses of welded sandwich steel plates are now described [33].

(1) Fatigue Tests of Sandwich Steel Weldments

The sandwich steel plate, composed of two 5 mm thick structural steel plates with synthetic resin in the middle, was specially developed to reduce the structural borne noise problem. Because of the thin layer of resin (about 0.3 mm), the loss factor of the sandwich plate is 100 to 200 times more than that of structural steel.

Five types of welded joints, as shown in Fig.4.5, were fabricated and fatigue tested;
a) welded joint with in-plane side gussets (GS-type),
b) welded joint with out-of-plane side gussets (TS-type),
c) welded joint of out-of-plane gussets (MS-type)
d) welded joints of transverse groove weld (CS-type),

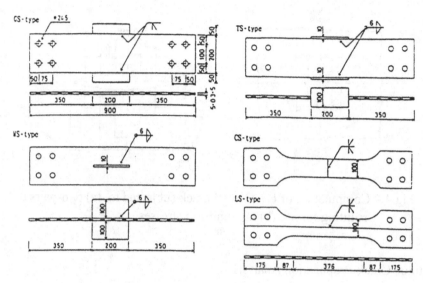

Fig.4.5 Fatigue test specimens of sandwich steel welded joints.

e) longitudinally groove-welded joint (LS-type).
Typical fatigue propagation patterns are shown in Fig.4.6, where fatigue fracture surfaces with some beach marking are shown. Fatigue crack propagation behavior of GS and LS specimens is discusse in the following paragraphs.

In the GS and the TS specimens, fatigue cracks initiated from the weld toe at the end of the gusset, and propagated perpendicularly to the applied axial load. After crack initiation, the fatigue crack propagated along the direction of the plate width as a quarter or half elliptical-shaped crack, until it penetrated through the plate thickness. The crack then propagated in the plate width direction. The LS specimens had small blowholes at the root near the synthetic resin layer, and fatigue cracks initiated from the blowholes. Cracks then propagated in the weld metal in the cross section perpendicular to the axial load, with a circular shape.

(2) Fracture Mechanics Analysis of Fatigue Crack Propagation Life
Based on phractographic observations, a fatigue crack growth model for each type of specimen was developed, and fracture mechanics analyses were carried out to compute the fatigue crack propagation life, N_p. When the fatigue crack initiates from the weld toe, an initial crack size of a_i=0.1 mm is assumed, and when the fatigue crack initiates from a blowhole, the value of a_i is taken as radius of the blowhole.

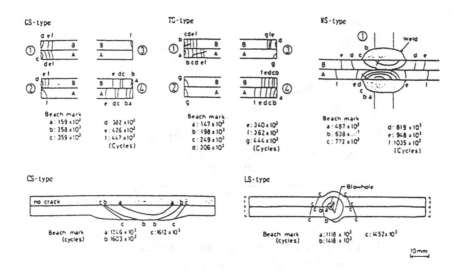

Fig.4.6 Fatigue fracture surfaces with beach marks.

For welded joints with in-plane side gussets (GS-Series), fatigue cracks normally initiated from the weld toe with a semi-elliptical shape. After the crack penetrated through the plate thickness, it propagated along the plate width direction. In sandwich steel plates, fatigue cracks initiated and propagated from one of the two steel plates with a quarter elliptical or semi-elliptical shape. After penetrating the plate thickness (5 mm), the crack propagated in the same way as for a normal plate.

The stress concentration factor K_t is first computed using a 2-D finite element method. It is assumed that the leg length is 6 mm, the flank angle of the weld toe is θ =45° and that the weld toe radius is ρ =0 mm. The geometry correction factor F_G is then computed. The crack shape is assumed to be quarter elliptical for the sandwich steel plate and semi-elliptical for the normal steel plate. The final crack size af is taken as 10 mm.

The computed N_p is plotted in Fig.4.7, along with the test results. The difference in N_p between the two plate types is small. This is because the only difference in the crack propagation behavior is due to the crack shape at the first stage of crack propagation, and the crack propagates faster at this stage due to the high stress concentrations at the weld toe. The computed N_p predicts the lower bound of the test data for the sandwich plate.

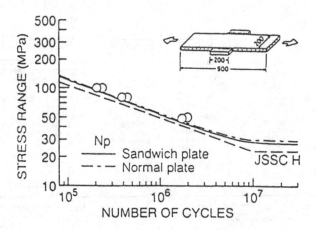

Fig.4.7 Comparison between test results and the analytical
propagation life of 200 mm gussets.

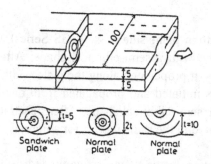

Fig.4.8 Analytical model of longitudinal groove welded specimens.

Consideration is now given to longitudinal groove welded joints (LS-Series). In
normal steel plates with longitudinal groove welds, fatigue cracks initiated either from
initial weld defects, such as blowholes, and then propagated with a circular shape, or
from the rough weld surface and then propagated in the direction of the plate thickness
with a semi-circular shape, as shown in Fig.4.8. In sandwich plates, fatigue cracks
originated from the inherent blowholes and propagated with a circular shape.
Blowholes were exposed by cooling the specimens with liquid nitrogen and breaking
the weld part. No fatigue crack initiated from the blowholes away from the fracture
surface.

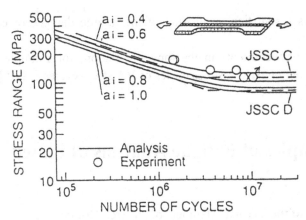

Fig.4.9 Comparison between test results and the analytical
crack propagation life model, for the various sizes of blowholes.

Values of N_p were computed for initial crack sizes a_i = 0.4, 0.6, 0.8 and 1.0 mm, and are plotted in Fig.4.9 along with the test data. For normal steel plates, the strength category is JSSC C, if the weld is of good quality, and JSSC D, if no specific NDT (non-destructive-test) is carried out to assure the weld quality. For the test specimens of sandwich steel plates, the equivalent crack sizes were between 0.5 and 0.7 mm. Since the computed N_p for a_i = 0.8 mm is longer than that of JSSC D, it can be applied to this type of detail. If any special treatment is possible to reduce blowholes, an even higher strength category for fatigue can be used.

4.5 Summary

A fracture mechanics analysis of the fatigue crack propagation life for weldments is summarized, and the applicability of this procedure is demonstrated for fatigue behavior of sandwich steel plate weldments. The fracture mechanics analysis on fatigue crack propagation life seems to have a wide range of applications for either test specimens or actual structures.

Some examples of possible applications are as follows:
a) The influence of residual stress, either tensile or compressive, can be evaluated quantitatively by considering the effectiveness of the stress intensity factor range. The simplest way may be to only consider the positive value of K in Δ K as being effective. A more complicated, but also more accurate definition is that Δ K is effective when the crack is fully open. This is called the effective stress intensity factor range, ΔK_{eff}.
b) Scatter in the fatigue life of welded joints can be evaluated by considering the scatter of weld shapes, weld toe irregularities, crack shapes, initial crack sizes, etc.

[29,35]. This enables a statistical evaluation of the fatigue damage of existing steel plated structures.

c) Fatigue behavior of weldments in the long life region, for example over 5×10^6 cycles, can be evaluated under CA stress cycles or VA stress cycles [34,35].

5. Some Examples of Fatigue Analysis of Plated Structures

5.1 Fatigue Analysis of the Vertical Members of an Arch Bridge

(1) Fatigue Cracks and Analysis

Bridge A, shown in Fig.5.1, was constructed in 1963 [36,37]. It is a two-hinged deck type arch bridge, with box girder arch ribs. The overall span length is 106.5 m, and the arch span and its rise are 85.0 m and 13.0 m respectively. The bridge is symmetric about its center line. After aproximately 20 years of service, fatigue cracks were observed at the upper and/or lower ends of the vertical members, as indicated by circles in Fig.5.1. Most of the cracks occurred in the shorter vertical members, and their distribution was symmetric with respect to both the transverse and longitudinal axes of the bridge.

Structural analyses were carried out for the original structure and for a retrofitted structure under moving load. Because of symmetry of the box girder the structure was modeled as a plain frame as shown in Fig.5.2. To study the influence of joint conditions on the stress distribution, analyses were carried out for a variety of cases. In addition to this, different cross sections for the vertical members, as well as the influence of bracing were considered for the proposed rehabilitation schemes.

The overall analysis schemes were as follows.

(a) Model A-1 considered the original structure, assuming the joints between vertical members and arch rib or girder to be fixed.

(b) Model A-2 considered the original structure, assuming the joints between vertical members and the arch rib or girder to be hinged.

(c) Model A-3 considered a revised structure, with increased cross sections for the short vertical members V7. Model A-4 considered a revised structure with increased cross sections for the vertical members V4 through V7. Model A-5 considered a revised structure with increased cross sections for the vertical members V2 through V7, to give the same cross section as vertical member V1.

(d)Model A-6 considered a revised structure with diagonal bracing of the vertical

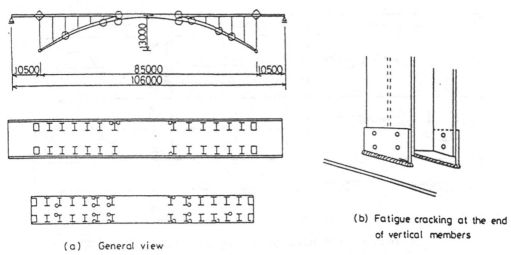

(a)   General view

(b) Fatigue cracking at the end of vertical members

Fig.5.1 Plan and elevation of 2-hinged arch bridge(Bridge A).

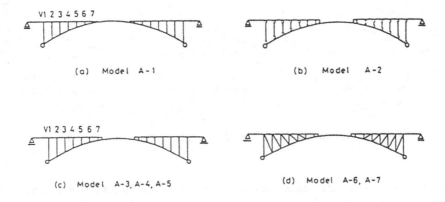

(a)   Model A-1

(b)   Model   A-2

(c)   Model   A-3, A-4, A-5

(d)   Model   A-6, A-7

Fig.5.2 Analytical model of fatigue damaged and stiffened arch bride.

members. End connections were assumed to be hinged. Model A-7 considered a revised structure with diagonal bracing of the vertical members, and end connections assumed to be fixed.

Fig.5.3 shows the deformed shape of the arch and stress waves of members V2 and V6, due to a moving truck load of 20 tonf. Analytical results show an increase in the bending stiffness of the short vertical members decreases the deformation and the stress range, and thus prolongs the fatigue life. It is also advisable to use fatigue resistant structural details at the ends of the vertical members to avoid any further

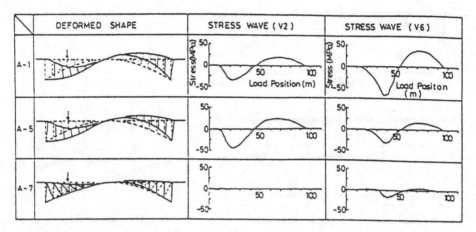

Fig.5.3 Comparison of deformations and stresses given by analyses.

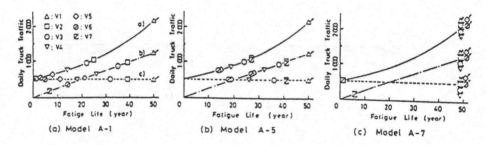

Fig.5.4 Estimated fatigue life of the ends of vertical members.

fatigue cracks. Significant reductions in both deformations and stress ranges are obtained when diagonal members are added, as is demonstrated by Model A-7.

(2) Estimation of Fatigue Life

The ends of the vertical members were fillet welded directly to the upper flange of the arch ribs or girders, as shown in Fig.5.4. Fatigue cracks formed either in the arch rib at the end of the gusset weld toes, or in the gussets themselves along the fillet weld toe. This joint is assumed to be a load carrying fillet weld, so joint classification E(80) of JSSC [12] is applied, assuming that the fatigue crack emanates from the fillet weld toe. The bridge is situated in a rural area, and to estimate the fatigue life, three types of traffic conditions were assumed, as shown in Fig.5.4. These conceptions assumed that either; (a) the number of daily trucks was initially 500, and increased by 3% per annum, (b) the number of trucks increased constantly by 250 per annum, or (c) it was 500, and remain constant. The weight of the trucks was taken as 20 tonf.

The computed fatigue life of each vertical member is plotted for each assumed traffic condition in Fig.5.4. For the original structure (Model A-1) the short vertical members, such as V7, V6 and V5, show relatively short fatigue lives, which corresponds with observations made of bridge A. When the stiffness of the vertical members is increased, such as Model A-5, the fatigue lives also increase. When future truck traffic is expected to decrease, this rehabilitation measure, in addition to fatigue resistant details at the ends of the vertical members, may be sufficient. When diagonal bracing members are added, all vertical members show sufficient fatigue resistance, except for V7 because diagonal braces can not be added near this member due to lack of space. This rehabilitation measure seems to be the most favorable way of increasing the fatigue resistance of the vertical members. Unfortunately, it requires expensive repair work.

5.2 Diaphragm Corners in a Box Girder Bridge

(1) Model of Actual Bridges

Analyses have also been carried out for diaphragm corners in a box girder bridge, as shown in Fig.5.5 [38]. Fatigue cracks were observed at diaphragm corners after about 20 years of service. Corners which exhibited fatigue cracks were repaired and strengthened using splice plates, which were attached by friction grip high strength bolts. Additional braces were also provided at sections with diaphragms, and lateral ribs were added in order to strengthen the box sections themselves. In this bridge, four lateral ribs are located at 1540 mm centers, between diaphragms of 7700 mm centers.

In order to analyze the corner of a diaphragm, part of the corner was modelled as shown in Fig.5.6. A load was applied at the end of the lateral rib, so that the corner was subjected to tensile stress. Model M01 represents a corner on the inside of the actual box girder bridge. When the cantilever part was added it became model M11. Model M02 allowed for stiffeners welded to the web of the floor rib, at the corner of the diaphragm. These stiffeners smoothly transfered the force in the flange of the vertical rib to the web of the beam. Model M03 represented the rehabilitated structure after repair works were carried out. A truss was formed inside of the box section to stiffen the box section. Model M13 allowed for diagonal members both inside and outside of the box girder. The diagonal members outside of the box supported the cantilever parts.

(2) Analytical Results

Fatigue cracks in the actual bridge initiated from the weld toe of the lower flange of the lateral rib at the diaphragm corner. The cracks penetrated through the flange plate, and propagated along the weld beads between the web and the flange. Computed stress distributions of σ_x and σ_y at the lower surface of the lateral rib,

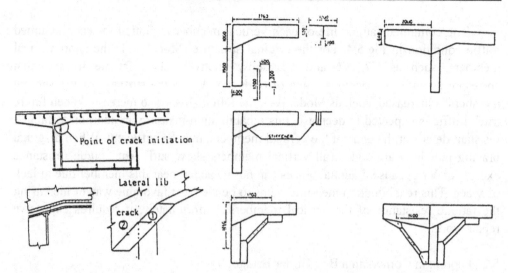

Fig.5.5 Fatigue crack found at the corner   Fig.5.6 FEM analytical model of
        of a diaphragm of a box girder bridge.   diaphragm corner.

along the lower flange, are shown in Fig.5.7. The x-axis shows the longitudinal
direction of the lateral rib, and the y-axis shows direction perpendicular to the lateral
rib. The origin of the abscissa is the flange corner in the box girder.

High stress concentrations can be observed at the corner of model M01. For
M02, which modeled stiffeners attached to the web of the lateral rib, decreased by
about 30%, and decreased by about 50%. When the diagonal members were allowed
for model M03, the stresses decreased by aproximately 81%, compared with model
M01. Forces passed mainly through the braces, so the reduction in stresses at the
corner was significant.

In order to investigate the plate bending effect, stresses at the upper and lower
surfaces of the lower flange are shown in Fig.5.8. The plate bending effect diminished
as distance from the corner increased, and the stress distribution became the same as
for σ_x in Fig.5.7. In the lower flange of the lateral rib for model M02. Plate bending
was less than half of that for the model M01 at the corner. For M03, the plate bending
effect in the flange was again reduced. Models M11 and M13 showed approximately
the same stress distributions as models M01 and M03.

(3) Analysis of Fatigue Crack Propagation
Fatigue cracks at the corner of diaphragms were assumed to initiate from weld
toes and to propagate through flange plates. In the analysis, an initial crack size of $a_0 =$
0.2 mm, final crack size of a_f=10 mm, and a semi-elliptical crack shape with a/b=1/2

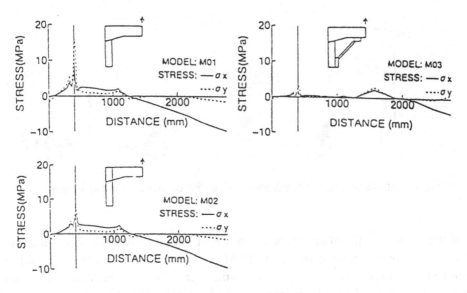

Fig.5.7 Computed stress distributions.

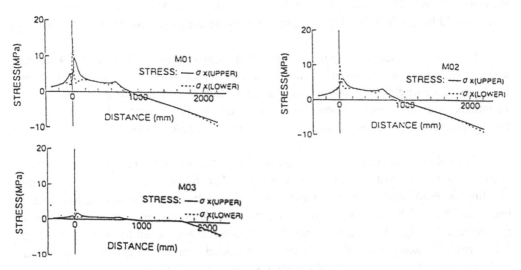

Fig.5.8 Stress due to plate bending.

were assumed.

Stress measurements were carried out using a weighed test truck of 37 tons before and after the repair and rehabilitation works. Several strain gages were attached

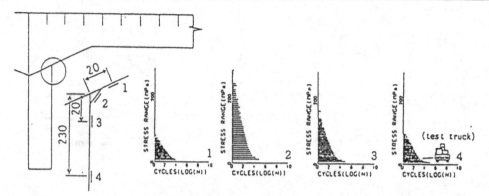

Fig.5.9 Stress histogram measurements at the corner of the diaphragm.

around the corners of the diaphragms, as shown in Fig. 5.9. By comparing measured stresses with those computed using the FEM analysis, stresses at corners could be determined by extrapolation. For example, strain gage No. 4 was located 230 mm away from the corner. The computed stress was 5.0 MPa at the gage position, and 8.0 MPa at the corner. The ratio between the two is 1.6. Under the test truck, the measured stress at the position of gage No.4 was 16.5 MPa. Therefore, the stress at the corner could be estimated as 1.6 times of the stress at the gage position, i.e. 26.5 MPa.

Service stress range histograms were also measured over a period of 24 hours using a histogram recorder. They are shown in Fig.5.9 for various measurement locations. It can be seen that there is a large scatter of service stress ranges due to variations in truck weight. Some trucks were obviously over-loaded, and/or heavy trucks appeared simultaneously and frequently in two or three lanes.

The value N_p is determined by solving the first order differential equation which is obtained by substituting Eq.4.1 into Eq.4.2. The life in years is determined by dividing the computed N_p by the number of cycles in each year. Computed lives were between 0.8 and 3.8 years, which is shorter than the observed period of about 20 years for the actual bridge. Since strain gage No.2 was directly attached to the fillet weld at the corner, it was assumed that stresses included the influence of local stress concentrations. When the effect of the local stress concentrations was neglected in the analysis, the computed fatigue life based on the stresses at strain gage No.2 became 14.5 years, which is closer to what was observed for the actual bridge.

After the bridge was strengthened by adding diagonal members, in a similar way as for models M03 and M13, stress measurements were again carried out. The stress at the corner of the diaphragm was estimated as 4.5 MPa when the weighed truck passed over the bridge. This is about one sixth of the stress before the rehabilitation works took place. If the pattern of daily truck traffic is assumed to stay the same as when the stress histogram shown in Fig.5.9 was measured, the computed fatigue life becomes over 600 years. This implies that the rehabilitation works are sufficient.

(4) Fatigue Life Considering the Plate Bending Effect

N_p was re-computed considering the local plate bending effect of the flange of the lateral rib. Firstly, the tensile stress component applied at the corner was determined considering the hot spot stress, which was obtained from the stress distribution at the center of the plate thickness using the FEM stress analysis. Then, the bending component was determined from the stresses at the upper and lower surfaces of the plate, as shown in Fig.5.7.

Since the bending stress component was large at the lower surface of the flange, where fatigue cracks initiated, and small at the upper surface, the stress decreased as the fatigue cracks grew. Fatigue cracks grew faster in the early stage of crack propagation. As a result, the fatigue crack propagation life was longer than that determined from the stresses neglecting the plate bending effect. Except for strain gage No.2, which was attached to the fillet welds at the corner, the computed fatigue life was between 14.5 and 30.6 years. This agrees well with the actual structure, since fatigue cracks were actually observed after about 20 years of service.

5.3 Summary

Two examples of fatigue analysis are demonstrated in this chapter. The First example considers the analysis of the fatigue cracked ends of the vertical members of a two-hinged arch bridge. Stress analysis was carried out using the finite element method and the fatigue life was estimated by comparing design S-N curves.

The second example considers a crack emanating from a fillet weld at the corner of a diaphragm of a box girder. Stress analysis was carried out for a simple model, and stresses compared with measured values. Fatigue crack propagation analysis was carried out for the estimated stress histograms at the corner, considering the plate bending effect. Results are compared with the observation of fatigue cracks in the ctual box girder bridge.

References

1.Fisher,J.W.: Fatigue and Fracture in Steel Bridges, Case Studies, John Wiley and Sons, Inc., New York, 1984.

2. Naess,A.A.(ed): Fatigue Handbook-Offshore Steel Structures, Tapir, Trondheim, 1985.

3. Gurney,T.R.: Fatigue of Welded Structures, Cambridge University Press,Cambridge, 1979.

4. Maddox,S.J.:Fatigue Strength of Welded Structures, Abington Publisher,Cambridge, 1991.

5. Yamada,K., Sakai,Y. and Kikuchi,Y.: Fatigue of tensile plate with gussets and stop holes as crack arrest, Proc. of JSCE,341(1984), 129-136.(in Japanese)

6. Yamada,K.et al.: Fatigue strength of tension member with welded gussets and life estimation by fracture mechanics, IIW Doc. XIII-1204-84,(1984).

7. Yamada,K.et al: Fatigue strength of two-and four year weathered weldments of weathering steels and structural steel, Proc. of JSCE, 337(1983),67-74.(in Japanese)

8. Yamada,K. and Kikuchi, Y.: Fatigue tests of weathered welded joints, Journal of structural Engrg., ASCE, Vol.110,No.9 (1984), 2164-2177.

9. Sakamaki, K. and Yamada, K. : Fatigue data base for welded joint, Proc. of JSCE, 356(1985), 547-553.(in Japanese)

10. BS5400: Steel, concrete and composite bridges, Part 10.Code of practice for fatigue, British Standard Institution (1980).

11. ECCS TC-6, Fatigue: Recommendations for the Fatigue Design of Steel Structures, ECCS, No.43 (1985).

12. JSSC: Recommendations for Fatigue Design of Steel Structures, Gihodo, Tokyo 1993.(in Japanese)

13. Miki,C. et al.: Fatigue of large-sized longitudinal butt welds with partial penetration, Proc. of JSCE, 322 (1982), 143-156.

14. Tajima,J. et al.: Fatigue strength of truss made of high strength steels, Proc. of JSCE, 341 (1984), 1-10.

15. Shimokawa,H.et al.: Effects of stress ratios on the fatigue strengths of cruciform

fillet welded joints, Proc. of JSCE, 344 (1984), 121-128.

16. Shimokawa,H. et al.: A fatigue test on the full-size truss chord, Proc. of JSCE, 344 (1984), 95-102.

17. Sakamoto,K. et al.: An investigation on fatigue strength for transverse chord of stiffening truss, Proc. of JSCE, Structural Eng./ Earthquake Eng., Vol.1, No.2, (1984), 157-167.

18. Simokawa,H. et al.: Fatigue strengths of large-size gusset joints of 800MPa class steel, Proc. of JSCE, Structural Eng./ Earthquake Eng., Vol.2, No.1 (1985), 279 -287.

19. Sakai,Y. et al.: Fatigue behavior of groove weld repair of cracked beams, Journal of Structural Eng., JSCE, 33A (1987). (in Japanese)

20. Yamada,K. et al.: Weld repair of cracked beams and residual fatigue life, Proc. of JSCE, Structural Eng./ Earthquake Eng., Vol.3, No.2 (1986),373-382.

21. Yamada,K. et al.: Fatigue strength of tension members with welded gussets and life estimation by fracture mechanics, IIW Doc XIII-1204-86(1986).

22. Yamada,K.: Fatigue of tension members with welded gussets and repair of cracks, Proc. of International Conference on Fatigue of Welded Construction, Brighton (1987).

23. Rolfe,S.T. and Barsom,J.M. : Fracture and Fatigue Control in structures Application of Fracture Mechanics, Prentice-Hall, New Jersey, 1977.

24. National Research Institute for Metals(NRIM): Fatigue Crack Propagation Properties in Arc-welded Butt-Joints of High Strength Steel for Welded Structure, NRIM Fatigue Data Sheet Technical Document No.3, Tokyo 1984.

25. Tada,H. and Paris,P. and Irwin,G.R.: The Stress Analysis of Cracks Handbook, Del Research Corporation,1973.

26. Albrecht,P. and Yamada,K.: Rapid calculation of stress Intensity factor, Proc.of ASCE, Vol.103, No.ST2(1977), 377-389.

27. Yamada,K. and Hirt,M.A.: Parametric fatigue crack propagation from fillet weld toes, J. of Structural Division, Proc. of ASCE, Vol.108, No.ST7, (1982), 1526

-1540.

28. Yamada,K. and Hirt,M.A: Parametric fatigue analysis of weldments using fracture mechanics, Proc. of JSCE, 319(1982), 55-64.

29. Yamada,K. and Nagatsu,S.: Evaluation of scatter in fatigue life of welded details using fracture mechanics, Proc. of JSCE, Structural Eng./Earthquake Eng., Vol.6, No.1(1989),13-21.

30. Yamada,K. and Hirt,M.A.: Fatigue life estimation using fracture mechanics, IABSE Colloquium, Lausanne(1982), 361-368.(in Japanese)

31. Yamada,K. et al.: Fatigue tests of welded details on the long life region and fracture mechanics analysis, IIW Doc. XIII-1281-88(1988).

32. Yamada,K. et al.: Fatigue analysis based on crack growth from toe of gusset end weld, Proc. of JSCE, 303(1980), 31-41.

33. Yamada,K. et al.: Fatigue behavior of welded joints of thick sandwich steel plates, Proc. of JSCE, 489/I-27(1994), 147-156.(in Japanese)

34. Yamada,K. and Shigetomi,H.: Fatigue tests of welded details in long life region and fracture mechanics analysis, Proc. of JSCE, No.404/I-11(1989).

35. Yamada,K. and Cheng,X.H.: Fatigue life analysis on welded joints under various spectrum loadings, Journal of Structural Engineering, JSCE, Vol.69A(1983).

36. Yamada,K., Tsuchihashi,M. and Kondo,A.: On repair of fatigue affected bridges, Constructional Steel Design, World Developments, (1992),649-658.

37. Mizuki,A. et al.: Field measurement and repair of road arch bridges which have fatigue crack, Kawada Technical Report, 1985, 244-250(in Japanese).

38. Yamada,K. et al.: Analysis of welded details prone to fatigue cracking, Proc of 4th EASEC, Seoul(1993), 1333-1338.

ULTIMATE LIMIT STATES OF PLATE- AND BOX-GIRDERS

R. Maquoi
University of Liège, Liège, Belgium

SUMMARY

The design of steel plated structures (for buildings and possibly for bridges) is presently based on the philosophy of limit states. This series of six lectures is aimed at giving the basic concepts on which such an ultimate limit state design for plate- and box-girders is founded.

The two first lectures are devoted to examine the behaviour of plate elements. In Lecture 1, the plate element is considered as an ideal structure made of an indefinitely elastic material and the concept of elastic critical plate buckling load is introduced. In contrast, the actual behaviour is briefly examined in Lecture 2, where the concept of ultimate load is opposed to the elastic critical buckling load.

The design criteria for flanges and webs are the scope of Lecture 3. They are examined at a general viewpoint; more detailed information regarding the resistance to direct and shear stresses is given in Lectures 4 to 6.

The tension field models for webs subject to shear, which exploit the postbuckling strength reserve, are reviewed in Lecture 4. A special emphasis is put on the tension band model and on the simple postcritical model, which are both the basis of the present specifications of Eurocode 3 for buildings.

The resistance to direct stresses, which is determined based on the concept of effective width, is the subject of Lecture 5.

Last, Lecture 6 is devoted to stiffened compression flanges of box-girders. Some specific aspects are highlighted.

These lectures are aimed at delivering a scientific background rather than design specifications. The reader is begged to use and exploit the material accordingly.

René MAQUOI

LECTURE 1

BEHAVIOUR OF PLATE ELEMENTS
PART 1 : ELASTIC CRITICAL PLATE BUCKLING LOADS

1. INTRODUCTION

Plate buckling is an instability phenomenon which is specific to plates subject to in-plane loads. It is to thin plates what *column buckling* is to slender struts. It is likely to occur when a slender plate is directly or undirectly subject to *compressive stresses*.

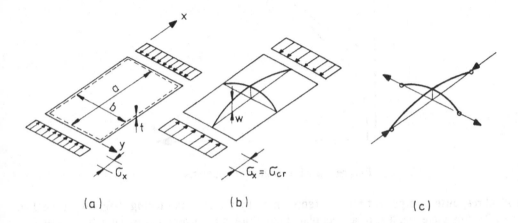

Figure 1.1 - Buckling of an ideal plate

For instance, let us consider (fig. 1.1) a simply supported rectangular plate of dimensions $a \times b$ and aspect ratio $\alpha = a/b$ close to unity, that is subject to uniaxial uniform direct compression. The plate of constant thickness t is assumed to be perfectly flat, made of an isotropic elastic material without residual stresses and loaded exactly in its middle plane. For small magnitudes of compressive stresses σ_x, the plate remains flat and any imposed transverse deflection w will disappear as soon as its cause is removed. For a specified magnitude of the compression load, $\sigma_x = \sigma_{x,cr}$, the plate will keep the deflected configuration produced by the imposed perturbation when the cause of this latter is removed. This magnitude is termed *elastic critical plate buckling stress*; it characterizes a state of neutral equilibrium. Such a plate buckles by *bifurcation of equilibrium*: indeed the plate remains flat for compression loads lower than the elastic critical plate buckling load while it is prone to get a deflected equilibrium configuration as soon as this limit load is reached.

In the buckled configuration, the fibres in the direction of compression have shortened because of elastic compressive strain, on the one hand, and bow effect in the longitudinal direction, on the other hand. This bow effect results in fibre lengthening in the transverse direction; the latter gives rise to *membrane effects* which have a stabilizing action and contribute a further increase in strength ability. Because this favourable contribution develops once the plate has buckled, it is termed *postbuckling strength*. The more the membrane action is allowed for - as a consequence of edge in-plane restraints -, the larger the postbuckling strength.

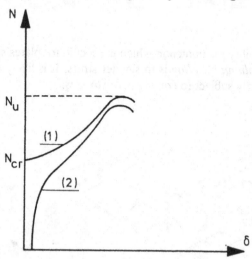

Figure 1.2 - Load-shortening curve

The ideal response of a perfect plate is represented by a load - shortening diagram (curve 1 in figure 1.2). Any plate element of a steel plated structure is fabricated industrially; by nature, it is affected by imperfections, more especially an initial out-of-flatness and residual stresses due to the manufacturing process. The behaviour of an imperfect plate is therefore somewhat different from the one of a perfect plate; indeed the initial out-of-flatness forces the plate to increase its deflection as soon as the load is applied (curve 2 in figure 1.2). For usual magnitudes of such an out-of-flatness, curve 2 approaches curve 1 when the load grows up. There is anyway an upper limit to the plate response. Material yielding results indeed in a decrease in the plate stiffness - yielded fibres do not contribute furthermore the plate flexural stiffness - and in a corresponding increase in the rate of the out-of-plane deflection; once first yielding occurs, the response curves progressively flatten till they reach a maximum load N_u, termed *collapse* (or *ultimate*) *load*, for which the plate stiffness has vanished (fig. 1.2). Residual stresses, due to the manufacturing process, cause prematurous yielding and precipitate the drop in plate stiffness. A real plate buckles by *divergence of the equilibrium* and exhibits a pre- and post- buckling non linear behaviour.

Above described behaviour is not specific to uniaxial uniform compression; it is observed in any plate subject to direct compressive stresses or to principal compressive stresses (when pure shear for instance).

The *elastic critical plate buckling stress* of an unstiffened plate may be basically derived from the well-known differential equilibrium equation which governs the linear theory of plate buckling :

$$\nabla^2\nabla^2 w \equiv \frac{\partial^4 w}{\partial x^4} + 2\frac{\partial^4 w}{\partial x^2 \partial y^2} + \frac{\partial^4 w}{\partial y^4}$$
$$= \frac{t}{D}\left[\sigma_x \frac{\partial^2 (w+w_o)}{\partial x^2} + 2\tau_{xy}\frac{\partial^2 (w+w_o)}{\partial x \partial y} + \sigma_y \frac{\partial^2 (w+w_o)}{\partial y^2}\right] \tag{1.1.}$$

where $D = Et^3/12(1 - v^2)$ is the flexural stiffness of the plate, t the thickness, E the Young modulus and v the Poisson ratio; $w(x,y)$ is the out-of-plane deflection of the plate. Basic loading cases are normally investigated by considering separately each load component in the second member of equation (1.1.). Effects of coincident loadings are obtained by means of interaction formulae.

The *pre- and postbuckling ranges* of such an imperfect plate can only be analysed by implementing above equation in view to account for the initial out-of-flatness and the membrane action ; as the latter produces stretching of the middle plane, both in-plane and out-of-plane behaviours are coupled. The solution of the non linear problem requires to solve a set of two coupled differential equations:

$$\nabla^2\nabla^2 w = \frac{t}{D}\left[\frac{\partial^2 \phi}{\partial x^2}\frac{\partial^2 (w+w_o)}{\partial y^2} - 2\frac{\partial^2 \phi}{\partial x \partial y}\frac{\partial^2 (w+w_o)}{\partial x \partial y} + \frac{\partial^2 \phi}{\partial y^2}\frac{\partial^2 (w+w_o)}{\partial x^2}\right] \tag{1.2.}$$

$$\nabla^2\nabla^2 \phi = E\left\{\left[\frac{\partial^2 (w+w_o)}{\partial x \partial y}\right]^2 + \frac{\partial^2 w_o}{\partial x^2}\frac{\partial^2 w_o}{\partial y^2} - (\frac{\partial^2 w_o}{\partial x \partial y})^2 - \frac{\partial^2 (w+w_o)}{\partial x^2}.\frac{\partial^2 (w+w_o)}{\partial y^2}\right\} \tag{1.3.}$$

where $\phi(x,y)$ is the Airy stress function and $w_o(x,y)$ the initial out-of-flatness. Equations (1.2.) and (1.3.) reflect respectively *equilibrium* and *compatibility* conditions.

Yet the integration of these equations is not at all easy in the elastic range. Consideration of the material non-linearities - material yielding and residual stresses - would add to the complexity of the mathematical treatment. The trend is to substitute physical models which are based on some idealizations and simplifications and allow for the assessment of the ultimate carrying capacity of plates subject to different types of loading. Such models are likely to reflect the influence of the governing parameters; most of the time, they must be calibrated against test results before being used as design models.

The critical plate buckling stress is a characteristic of the perfect plate, which is itself a limit case for the similar imperfect plate. It will therefore be a kind of reference parameter in any design process based on ultimate models. Accordingly it is worthwhile reminding briefly the reader with the main results of the *linear theory of plate buckling*.

2. ELASTIC CRITICAL PLATE BUCKLING STRESS

It is not the place here to review the different methods which may be contemplated with a view to derive expressions for plate buckling stresses. In the literature, the critical buckling stress σ_{cr}, or τ_{cr}, of a plate is usually given with reference to the buckling stress of a transverse pin-ended strip of unit width and length b, (fig. 1.1), termed *Euler reference stress* σ_E, according as:

$$\sigma_{cr} = k_\sigma \sigma_E \qquad (1.4.)$$
$$\tau_{cr} = k_\tau \sigma_E \qquad (1.5.)$$

where the reference stress writes :

$$\sigma_E = \pi^2 D / b^2 t$$
$$= \left[\pi^2 E / 12(1-v^2)\right](t/b)^2 \qquad (1.6.\text{a-b})$$

i.e. for steel $(v = 0.3)$:

$$\sigma_E \approx 0.9 E(t/b)^2 \approx 190.000(t/b)^2 \quad (in\ N/mm^2) \qquad (1.6.\text{c})$$

Dimensionless coefficients k_σ and k_τ are termed *buckling coefficients*; they depend on several parameters : the plate aspect ratio, the stress distribution, the boundary conditions and, for stiffened plates, the relative flexural, torsional and axial stiffness properties of the stiffeners.

3. PLATE SUBJECT TO UNIAXIAL UNIFORM COMPRESSION

For a long *simply supported* plate subject to *uniaxial uniform compression*, the buckling coefficient is :

$$k_\sigma = (m/\alpha + n^2\alpha/m)^2 \qquad (1.7.)$$

where m is the number of half buckling waves in the direction of the compression and n, the number of such waves in the transverse direction. Because only the smallest value of k_σ is worthy, n shall be taken equal to unity; that means that the plate buckles with one single half wave in the transverse direction. The relevant curve $k_\sigma = f(\alpha)$ is made of successive festoons, the lower envelope of which is solely significative (fig. 1.3). It appears thus that the buckling mode changes when the aspect ratio grows up. Each festoon corresponds to a buckling mode, i.e. to a given number of half waves in the direction of compression. The minimum of each festoon is reached when $m = \alpha$, in which case $k_{\sigma,min} = 4$; a useful and only slightly safe approximate consists in assuming $k_\sigma = 4$ for $\alpha \geq 1$.

Boundary conditions better than simple supports result in a higher buckling coefficient and therefore in a higher critical plate buckling stress. That is clearly reflected by comparing the curves A, B and C obtained for different boundary conditions (fig. 1.4). Most often the edges are neither simply supported nor fully clamped but well elastically restrained. Usually an accurate assessment of the degree of restraint is not practicable because the restraint is provided by connections and the remaining part of the structure. Therefore the conservative assumption of simple supports is usually made for the web of a fabricated I section. Other interesting cases are plates having one longitudinal free edge and the other longitudinal edge either simply supported or clamped (curves D and E in figure 1.4). The free/simply supported boundary conditions are especially worthy for the consideration of the flanges of an I section.

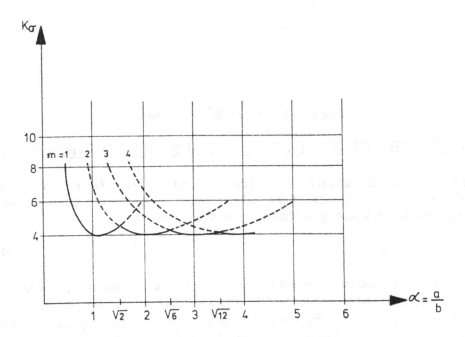

Figure 1.3 - Buckling coefficient for uniform compression (simple supports)

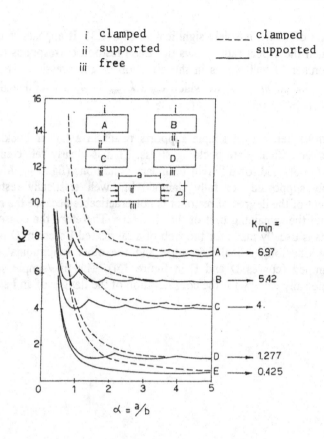

Figure 1.4 - Influence of boundary conditions

4. PLATE SUBJECT TO A LINEAR DIRECT STRESS DISTRIBUTION

When two opposite edges of a rectangular plate are subject to a *linear direct stress distribution* such that a portion or the whole of the width is subject to compressive direct stresses, the buckling coefficient is depending in addition on the *direct stress ratio* :

$$\psi = \sigma_2 / \sigma_1 \tag{1.8.}$$

where σ_2 and σ_1 are the minimum and maximum direct stresses respectively; thus $\psi = 1$ for uniform compression and $\psi = -1$ for pure bending. The chart obtained for pure bending $(\psi = -1)$ demonstrates that the buckling mode changes again when α grows up and that clamped longitudinal edges provide an increase of about 70 % in the buckling coefficient, compared to simply supported edges (fig. 1.5). For sake of simplicity, k_σ is usually taken as $k_{\sigma,min} \approx 40$ as soon as α exceeds 2/3 when simply supported edges. Such charts can be drawn for any value of ψ. Only the chart of $k_{\sigma,min}$ versus the stress ratio ψ (fig. 1.6) is given here.

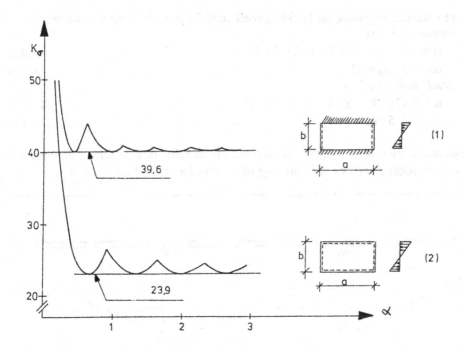

Figure 1.5 - Buckling coefficient for pure bending ($\psi = -1$)

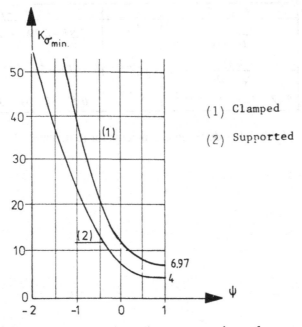

(1) Clamped

(2) Supported

Figure 1.6 - Influence of the direct stress ratio on $k_{\sigma,min}$

With the aim to practice purposes, the buckling coefficient for *pure bending* may write :
- for *simply supported edges* :

$$\alpha \leq 2/3 : k_\sigma = 15{,}87 + 1{,}87/\alpha^2 + 8{,}6\,\alpha^2 \qquad (1.9.a)$$

$$\alpha > 2/3 : k_\sigma = 23.9 \qquad (1.9.b)$$

- for *longitudinal clamped edges* :

$$\alpha \leq 0.475 : k_\sigma = 21{,}3 + 2/\alpha^2 + 42\,\alpha^2 \qquad (1.10.a)$$

$$\alpha > 0.475 : k_\sigma = 39.6 \qquad (1.10.b)$$

When plate elements are subject to direct stress distributions other than uniform compression or pure bending, the buckling coefficient k_σ for *long plates* may be drawn from Table 1.1.

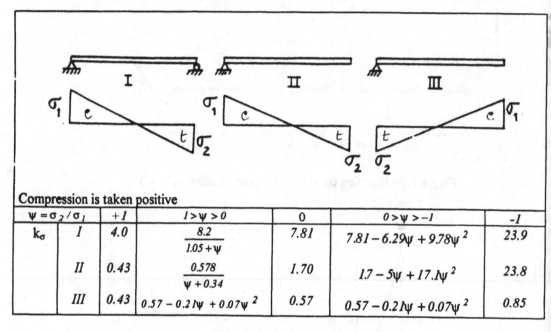

Compression is taken positive

$\psi = \sigma_2/\sigma_1$		$+1$	$1 > \psi > 0$	0	$0 > \psi > -1$	-1
k_σ	*I*	4.0	$\dfrac{8.2}{1.05 + \psi}$	7.81	$7.81 - 6.29\psi + 9.78\psi^2$	23.9
	II	0.43	$\dfrac{0.578}{\psi + 0.34}$	1.70	$1.7 - 5\psi + 17.1\psi^2$	23.8
	III	0.43	$0.57 - 0.21\psi + 0.07\psi^2$	0.57	$0.57 - 0.21\psi + 0.07\psi^2$	0.85

Table 1.1. - Buckling coefficients for long plates

5. PLATE SUBJECT TO PURE SHEAR

When a thin rectangular plate is subject to *pure shear*, it may also be prone to plate buckling because pure shear results in equal but opposite principal stresses sloped at ± 45° on the directions of pure shear. Principal compressive stresses may cause plate buckling while principal tensile stresses help to delay it. In contrast to most cases of loadings by direct stresses, the buckling coefficient curve for pure shear is monotoneously decreasing (fig. 1.7); it may be approached as follows :

- for all *simply supported edges* :

$$k_\tau = 4 + 5{,}34/\alpha^2 \quad (\alpha \leq 1) \tag{1.11.a}$$
$$ = 5{,}34 + 4/\alpha^2 \quad (\alpha \geq 1) \tag{1.11.b}$$

- for all *clamped edges* :

$$k_\tau = 5{,}6 + 8{,}98/\alpha^2 \quad (\alpha \leq 1) \tag{1.12.a}$$
$$ = 8{,}98 + 5{.}6/\alpha^2 \quad (\alpha \geq 1) \tag{1.12.b}$$

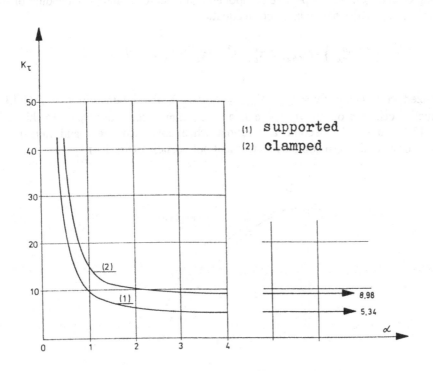

Figure 1.7 - Buckling coefficient for pure shear

6. PLATE SUBJECT TO COMBINED SHEAR AND DIRECT STRESSES

When a rectangular plate is subject to *combined shear and direct stresses,* the buckling resistance is influenced by the interaction between the load components. Let us use the following symbols :

a) $\sigma_{cr}^o = k_\sigma \, \sigma_E$ and $\tau_{cr}^o = k_\tau \, \sigma_E$: the individual critical stresses which should cause plate buckling, when acting *separately;*

b) σ_{cr} and τ_{cr} : the critical stresses which should produce buckling, when acting *coincidently.*

The stresses (a) may be computed based on above Sections 4 and 5. The stresses (b) are drawn from an interaction relation, which is just an engineering approach and is most often only an approximate. For instance, such an interaction formula is :

$$0.25(1+\psi)(\sigma_{cr}/\sigma_{cr}^o)+\sqrt{\left[0.25(3-\psi)(\sigma_{cr}/\sigma_{cr}^o)\right]^2+(\tau_{cr}/\tau_{cr}^o)^2}=1 \quad (1.13)$$

Designating σ_c and σ_b as the respective components of axial force and pure bending of the direct stress distribution, an alternative interaction equation is :

$$\left(\sigma_{c,cr}/\sigma_{c,cr}^o\right)+(\sigma_{b,cr}/\sigma_{b,cr}^o)^2+(\tau_{cr}/\tau_{cr}^o)^2=1 \qquad (1.14)$$

In normalized coordinates ($y=\sigma_{cr}/\sigma_{cr}^o, x=\tau_{cr}/\tau_{cr}^o$), the interaction curve (1.13.) is an ellipse which is centered on the vertical axis and is symmetrical with respect to this axis. Both curves (1.13.) and (1.14.) become a parabola when combined shear and uniaxial uniform compression or a circle when combined shear and pure bending (fig. 1.8).

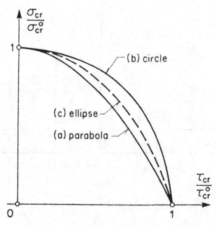

Figure 1.8 - Normalized interaction buckling curves

The real interaction curves depend on the aspect ratio; however above approximates, which do not depend on this factor, may be considered as quite satisfactory for practice purposes.

Of course both values σ_{cr} and τ_{cr} cannot be drawn from a single interaction equation. A second relation is necessary; therefore it is generally assumed that the loading is proportional, so that :

$$\sigma_{cr} / \tau_{cr} = \sigma / \tau = \bar{s} \qquad\qquad (1.15.)$$

The relevant ratio $\bar{s} = \sigma / \tau$ is presumably known for a reference state, for instance, in service conditions.

LECTURE 2

BEHAVIOUR OF PLATE ELEMENTS
PART 2 : ULTIMATE LOADS

1. INTRODUCTION

The *elastic critical plate buckling load* of a plate is not a realistic measure of the *carrying capacity*, i.e. of the *ultimate load*, because the plate is far from being perfect. Indeed the behaviour is affected, on the one hand, by unavoidable initial geometric and material imperfections, and, on the other hand, by the effects of inelasticity and of pre- and postbuckling stability.

The *geometric imperfections* consist mainly in an initial out-of-flatness of the plate; the latter produces a growth of out-of-plane deformations from the onset of loading, with the result of a loss in plate stiffness due to bow effect; however, at high level of strains, the behaviour is nearly not affected by the level of imperfection usually met in practice (fig. 2.1).

The onset of *inelasticity* is influenced not only by in-plane stresses but also by plate bending stress components. The larger the out-of-flatness, the earlier the yielding and thus the larger the effect of material yielding on the collapse mode. The consequence of inelasticity is a loss in plate stiffness too and a drop in collapse load, compared to a similar perfect elastic plate (fig. 2.2).

The *residual stresses* are the governing *material imperfections;* they precipitate the first yielding and thus contribute a further reduction in stiffness ; they may also affect the ultimate load to some degree.

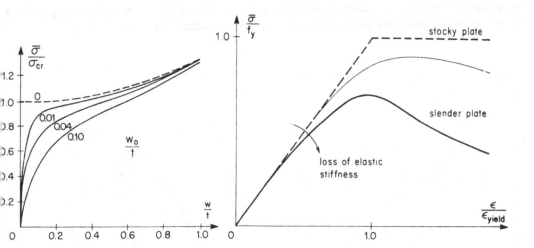

Figure 2.1 - Effect of out-of-flatness   Figure 2.2 - Effect of inelasticity

2. PLATE SUBJECT TO AXIAL COMPRESSION

Plates subject to *axial compression* have been the most extensively studied for what regards the elasto-plastic behaviour. Mostly numerical approaches have been used in this respect, the results of which have been compared with a lot of experimental results.

All these investigations were aimed at studying the effects of the main parameters on the stress-strain response of the plate. The governing parameter is the *normalized plate slenderness* $\bar{\lambda}_p$; it is defined quite similarly to the normalized column slenderness, i.e. :

$$\bar{\lambda}_p = \sqrt{f_y / \sigma_{cr}}$$
$$= (1.05 / \sqrt{k_\sigma})(b / t)\sqrt{f_y / E} \qquad (2.1.)$$

with $k_\sigma = 4$ for a long plate ($\alpha \gg 1$) subject to axial compression. The loss in strength due to both geometrical imperfections and residual stresses is relatively the largest in the vicinity of $\bar{\lambda}_p \approx 1$, where the interaction between yielding and plate buckling is the highest (fig. 2.3); that is the range of intermediate slenderness. On the other hand, the greatest loss in pre-collapse rigidity due to residual stresses is observed for plates of moderate and low slenderness ($b/t < 60$). Last, initial geometrical imperfections reduce generally the compressive strength and can produce a change in the failure shape from a gradual process to a more sudden buckling event with a significant subsequent loss in load carrying capacity, especially in the range of intermediate slenderness. It has been found that the minimum strength in axial compression is obtained for an aspect ratio close to unity. Long plates show a higher initial stiffness, a higher strength but a steeper unloading path after the maximum load has been reached (fig. 2.4).

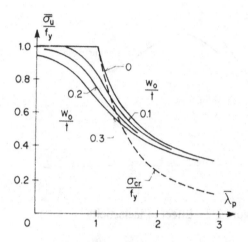

Figure 2.3 - Effect of plate slenderness
on the strength

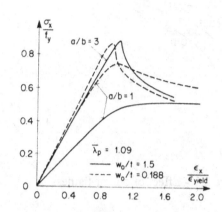

Figure 2.4 - Effect of plate aspect ratio
on the post-collapse behaviour

Regarding in-plane restraints, the unloaded edges are considered successively either as fully *restrained edges* - edge forces and non deformable boundaries -, or as *unrestrained edges* - free to pull in -, or as *constrained edges* - free from edge forces but remaining straight through symmetry-. No significant difference is observed between constrained and unrestrained edges when the plate is stocky ; in contrast, when slender plates, constrained strength approaches restrained strength (fig. 2.5). Restrained conditions provide the highest strength and unrestrained conditions the smallest strength.

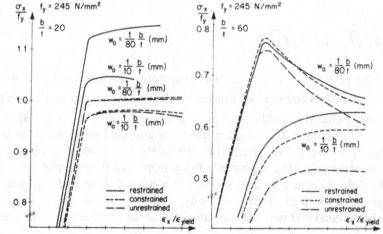

Figure 2.5 - Effect of in-plane restraints on strength and the post-collapse behaviour

Large initial out-of-flatness affect the strength whatever the plate slenderness; the larger the out-of-flatness, the smaller the ultimate load. The residual carrying capacity after unloading is nearly independent of the imperfection level in the range of intermediate and large plate slenderness values (fig. 2.6).

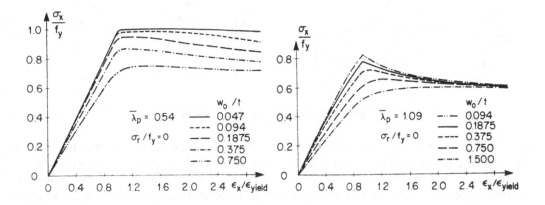

Figure 2.6 - Effect of plate out-of-flatness on the strength and the post-collapse behaviour

Numerical simulations demonstrate clearly that the effect of residual stresses is more marked for stocky plates, because of the larger interaction with overall yielding ; it tends to be masked by geometrical imperfections in slender plates. Thus the effect shall be the greatest in the medium range of plate slenderness values.

3. SHEAR

The behaviour of a plate subject to *pure shear* is not quite the same to the one of a plate subject to axial compression. Prior to buckling, the stress is mainly a combination of principal (diagonal) tensile and compressive stresses of equal magnitude. Once elastical critical shear plate buckling has occurred, the behaviour is still stable in the elastic regime because the loading can be resisted by an increase in the diagonal tensile force, provided the panel be surrounded by stiffening members onto which the membrane tensile stresses can anchor (fig. 2.7).

Here too, an initial out-of-flatness causes a growth of the out-of-plane deformations from the onset of loading. However, its effect is less marked for shear than for axial compression because of the presence of the tensile element of load resistance. For slender plates, deformations grow progressively and the ratio of diagonal tensile to compressive stresses gradually increases. Tensile component is imperfection sensitive; therefore it may dominate the behaviour to some extent and reduce the sensitivity of ultimate shear strength to out-of-flatness magnitude.

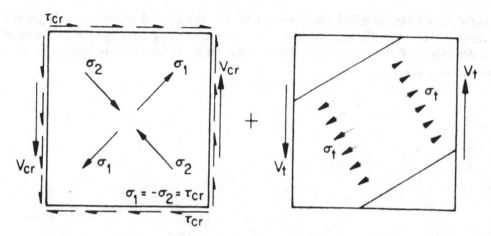

Figure 2.7 - Ultimate shear model

The effects of inelasticity and residual stresses on collapse are essentially similar to those observed for pure compression. Plastic shear resistance (yield shear load) is an upper bound for stocky panels. For slender panels, yielding at the outer fibres is a result of the growth in deformation; it contributes an increase in subsequent deformations and results in lower ultimate loads. It is while stressing that the yield stress limits the degree of tensile resistance that can be built up and therefore influences directly the contribution of the tension field action to the ultimate shear resistance.

More information regarding the assessment of the ultimate shear capacity is given in Lecture 4.

4. COMBINED SHEAR AND DIRECT STRESSES

Because the buckling modes corresponding respectively to the two load components are dissimilar, the imperfection sensitivity of the plate is between those produced by the basic stress components separately. That explains that a restrained panel subject to some proportions of axial compression and shear can exhibit a higher strength than the same panel subject to axial compression only (fig. 2.8).

Few numerical studies are available in this field. Nevertheless, they have produced interaction curves of ultimate stresses derived from sets of panel stress-strain responses. These curves are established for restrained and unrestrained panels, the direct stresses being uniform tensile or compressive in one direction only (fig. 2.8). The small imperfection sensitivity in the compression zone is quite observable, while, in the tension zone, the different curves tend quickly to merge because buckling becomes irrelevant.

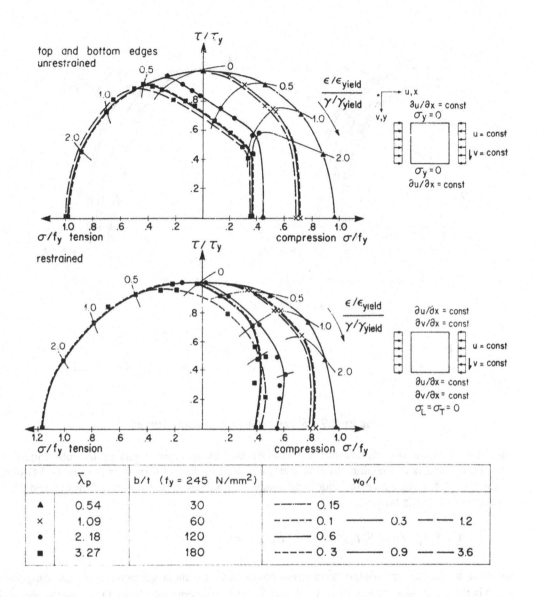

Figure 2.8 - Shear-compression interaction curves

Similarly interaction curves have been calculated for combined in-plane bending and shear; the moments are expressed in terms of full plastic resistance M_u or yield moment M_y (fig. 2.9). These curves exhibit a significant dependence of the strength on the level of imperfection for stocky plates, especially when unrestrained.

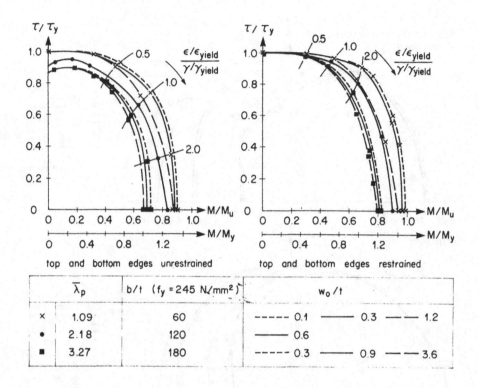

Figure 2.9 - Shear-bending interaction curves

The results of all the aforementioned studies and investigations enable to substantiate an interaction formula at the ultimate limit state, which has a similar format as the one (1.14) given in Lecture 1 for elastic critical buckling loads. For instance, for combined direct and shear stresses, this formula writes :

$$\sigma_c / S_c f_y + (\sigma_b / S_b f_y)^2 + (\tau\sqrt{3} / S_s f_y)^2 = 1 \qquad (2.2.)$$

where σ_c in the uniform compression stress component, σ_b the larger bending stress component and τ is the coincident shear stress. S_c, S_b and S_s are coefficients which, by factoring the material yield stress, provide the ultimate resistances to the individual loading components (fig. 2.10).

Formula (2.2.) does not account for strains. That means that the ultimate limit state is thus defined irrespective of the available ductility. A limitation of the maximum strain to 2,5 times the relevant yield strain should be appropriate.

A possible alternative approach for the interaction between bending and shear in plate - and box-girders is the tension field mechanism approach which is usually considered as a design approach. It is examined in Lecture 4.

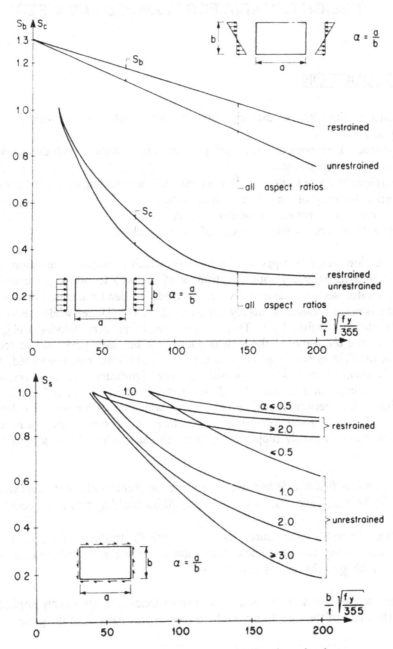

Figure 2.10 - Factors S_b, S_c and S_s versus the plate slenderness

LECTURE 3

DESIGN CRITERIA FOR FLANGES AND WEBS

1. INTRODUCTION

With regard to the ultimate carrying capacity under static or quasi-static loading, it is necessary to check :
a) the resistance of sections to *direct* and, possibly, *shear stresses* with due allowance made for their possible interaction;
b) the resistance to *local instability phenomena*, such as buckling of the compression flange and vertical buckling of the flanges into the web,
c) the *resistance to transverse concentrated forces*.
Member instability is not within the scope of present chapter.

The compression flange is supported by the web which provides, in addition, a rotational restraint. The latter is linked with the stiffness of the web to out-of-plane deflections. In welded fabricated sections, an intensive use is made of slender plates. For plate- and box girders, the bending moment is mainly experienced by the flanges while the web(s) is(are) mainly devoted to resist shear. That results usually in very slender webs; in bridge construction, the d/t_w ratio of the web is about 100 but may peak 200 in recent bridge projects; in the field of buildings, larger d/t_w ratios, up to 300, have been used. As a result, the web is normally not able to provide a very significant rotational restraint to the compression flange; in addition, should this restraint exist, it could be affected by the possible interaction between buckling phenomena which could appear coincidently in both the web and the compression flange. Therefore, it is commonly assumed that the compression flange is simply supported by the web plate, without taking advantage of any rotational restraint.

A slender web may first buckle because of in-plane longitudinal direct stresses which are due to longitudinal bending and, possibly, axial force. Shear buckling may also occur.

As a result of the bending curvature, transverse (vertical) stresses do develop at the web-to-flange junctions. They cause vertical compression in the web and result in a possible danger of vertical buckling of the flanges into the web.

Sometimes the web has also to resist concentrated loads that are mostly applied onto one flange, in the plane of the web, transversely to the longitudinal axis of the girder.

2. FLANGE RELATED CRITERIA

In a plate girder, the cross-sectional proportions - i.e. width and thickness - of the flanges are always small compared to the girder span; therefore shear lag cannot be significant. In addition the web depth of a plate girder is considerably larger than the flange thickness; it is thus quite acceptable to neglect the direct stress gradient across the flange thickness and to use, as reference flange stress, the stress at the flange centroïd. That corresponds to concentrate each flange at the level of its centroïd.

In a box girder, even the flanges are slender and sometimes the ratio of the flange width to the girder span is such that significant shear lag occurs. Similarly to plate buckling, *shear-lag* results in a non uniform distribution of direct stresses in fibres located at a same distance from the bending axis; however shear lag has nothing other in common with plate buckling. Shear-lag is a first-order effect. Its effects are not depending on the sign of the direct stresses, i.e. tension and compression. Thus, while plate buckling affects only the compression flange, shear may affect anyone of the compression and tensile flanges, when these flanges are wide compared to the span. The question of shear-lag is discussed more in deep in some other lectures.

As far as the steel plate elements can develop its full strength, the yield stress constitutes usually the limiting stress for the reference stresses in both tensile and compression flanges; by doing so, yielding restricted to half of the flange thickness (due to direct stress gradient across the flange thickness) is implicitly allowed for. However instability phenomena, when they occur in the elastic range, are likely to prevent the compression flange from reaching its full yield capacity. Among these phenomena, some of them are dealing with the overall buckling of the plate girder (column buckling, lateral torsional buckling) while some others are more concerned with local plate buckling. Only the latter are examined in the present lecture.

2.1. Tension flange

No instability can occur in the tension flange. It is presumably assumed that no shear-lag does affect the flanges of a plate girder. Therefore the limiting direct stress at the tensile flange centroïd is the material yield stress.

2.2. Compression flange of plate girders

The compression flange may buckle according to a buckling mode with which the web will in principle be concerned, because the latter is expected to provide the flange with an elastic restraint. It has yet been said that such a restraint should conservatively be fully neglected.

Normally the girder cross-section is symmetrical with respect to the web. Because of this symmetry, each half-flange of thickness t_f is idealized and modelled as a long uniformly compressed plate, the longitudinal edges of which are respectively simply supported and free (fig. 3.1).

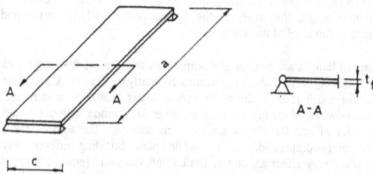

Figure 3.1 - Modelling of half the compression flange

Provided its aspect ratio a/c is large enough - about $a/c > 10$, what is mostly the case in practice -, it is well-known that the boundary conditions at the loaded edges are not at all determinative and that the buckling coefficient is $k_\sigma = 0.43$ for axial compression. The elastic critical plate buckling stress writes then :

$$\sigma_{cr} = 0.43 \, \sigma_E \tag{3.1.}$$

with :

$$\sigma_E = \left[\pi^2 E / 12(1-v^2)\right](t_f / c)^2 \tag{3.2.}$$

For an ideally perfect plate, plate buckling does not reduce the capacity of the plate as far as the critical plate buckling stress exceeds the yield stress of the flange; by introducing the general concept of normalized plate slenderness $\overline{\lambda}_p$:

$$\overline{\lambda}_p = \sqrt{f_y / \sigma_{cr}} \tag{3.3.}$$

above conclusion writes $\overline{\lambda}_p \leq 1$. Because a plate girder is especially a fabricated section, its components are far from being perfect; more especially, the compression parts of the section are never ideally flat and are subject to complex distributions of residual stresses. To account for both geometrical and structural imperfections in the compression flange, it is usually required that the normalized plate slenderness $\overline{\lambda}_p$ of this flange does not exceed a limiting value β lower than unity. Thus this condition for full efficiency writes :

$$c / t_f \leq 18{,}5 \, \beta \sqrt{235 / f_{yf}} \tag{3.4.}$$

where f_{yf} is the yield stress (in N/mm²) of the flange material. Looking at national codes, large variations in the value of β for full efficiency can be observed. For sake of consistency with Winter formula - which will be introduced later - β should be taken equal to 0.67 for sections of Class 3 and 4 and be reduced to 0.6 for sections of Class 1 and 2 because of some required plastic redistribution.

Of course, the flanges of a plate girder contribute the most the second moment of inertia about the strong bending axis of the section. There is thus an obvious benefit to proportion the width and thickness of the compression flange, so that the relevant width-to-thickness ratio complies with the full efficiency condition.

The limiting width-to-thickness ratio given by equation (3.4.) presumes that the yield stress in the flange is being reached. A less conservative value would be obtained by substituting possibly the design maximum stress σ_{max} (under factored loads) for f_{yf}. Should lateral torsional buckling be governing, the relevant ultimate stress σ_K will be the maximum stress in the compression flange. Because of lateral buckling, σ_K must be understood as the *average* stress in the flange, while the yield stress will be expectedly reached at the outwards tip of the flange. Due to the non uniform stress distribution over the flange width, it is recommended to substitute the geometric average stress $\sqrt{\sigma_K \cdot f_{yf}}$ for f_{yf} in (3.4.).

When the compression flange slenderness is such that a full efficiency cannot be reached, the determination of the section properties and of the relevant strength shall be conducted based on an *effective flange width* $2c_e$, where c_e is drawn from equation (3.4.) written as an equality.

3. WEB RELATED CRITERIA

Not only the web shall resist direct and shear stresses produced by longitudinal bending and possible axial force, but also the web shall also be designed to prevent *flange induced buckling* and possibly to resist *transverse concentrated forces*.

3.1. Flange induced buckling

In addition to provide shear resistance, the web of a plate girder is aimed at supporting - transversely - the flanges, which are generally subject, as a result of bending, to compression and tension respectively. The web stiffness which is required for this purpose, is not very large; however for plate girders with very slender web, the ultimate limit state may be governed by the *vertical buckling of the flanges into the web*; this phenomenon is never governing for hot-rolled sections. Such an ultimate limit state can be prevented by requiring that the web slenderness - i.e. the depth-to-thickness ratio d/t_w of the web - does not exceed a limiting value, which is determined hereafter.

In order to derive a rather simple but safe criterion, the reasoning is conducted on an unstiffened doubly symmetrical plate girder of constant depth and fully effective flanges. Then the neutral axis is located at mid-depth of the web so that, in pure bending of magnitude M, both flanges are subject to a same axial force $N = M/h$ (respectively in tension and in compression), where the distance h between the flange centroïds approaches the web depth d. As far as instability does not govern, the ultimate axial force N in the flanges is the squash load $A_f f_{yf}$, where A_f is the flange cross-sectional area. The curvature is constant; the equilibrium of the compression and tensile flanges respectively requires that uniformly distributed radial forces (fig. 3.2) be present; the latter result in *vertical compression in the web*. The vertical compressive stress writes :

$$\sigma_v = A_f f_{yf} / \rho t_w \qquad (3.5.)$$

where ρ is the radius of curvature.

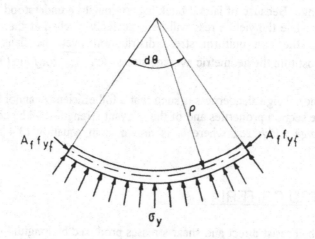

Figure 3.2 - Vertical stresses due to bending curvature

For sake of simplicity, the radii of curvature related to both flanges, are taken equal; both are indeed much larger than the web depth d. The radius of curvature ρ is related as follows to the strain ε_f in the flange :

$$\rho = d / 2\varepsilon_f \qquad (3.6.)$$

The strain ε_f must account for the residual tensile stress σ_r that must be overcome to achieve uniform yield in compression. Therefore, it writes for the compression flange :

$$\varepsilon_f = (f_{yf} + \sigma_r) / E \qquad (3.7.)$$

wherefrom :

$$\sigma_v = A_f f_{yf} (f_{yf} + \sigma_r) / 0.5 \, d \, t_w \, E \qquad (3.8.)$$

or, by introducing $\sigma_r = 0.5 f_{yf}$ (as a realistic assessment) and the web cross-sectional area $A_w = d t_w$:

$$\sigma_v = 3 (A_f / A_w) f_{yf}^2 / E \qquad (3.9.)$$

As the tension flange develops a similar, though slightly lower, vertical stress, the effect of bending curvature results in a vertical uniform compression of the web. Each vertical web strip of unit width acts as a column of inertia $t_w^3 / 12(1-v^2)$, that is conservatively assumed simply supported at both ends; its elastic critical buckling stress writes :

$$\sigma_{v,cr} = \left[\pi^2 E / 12(1-v^2)\right](t_w / d)^2 \qquad (3.10.)$$

It is generally accepted that, at flange failure, the safety factor against vertical buckling of the web may be taken equal to unity. That may be justified by the fact that the effect of web imperfections is more than offset by the conservatism of above assumptions and simplifications.

Therefore, the condition $\sigma_v \le \sigma_{v,cr}$ provides the d/t_w limiting ratio, which is required to prevent from vertical buckling of the flanges into the web :

$$d / t_w \le 0.55 \sqrt{A_w / A_f} (E / f_{yf}) \qquad (3.11.)$$

A similar result can be derived from a quite different approach. For this purpose, let us consider a vertical strip of unit width which is cut in the web; it is assumed to be pin-ended at its supports on the flanges and to exhibit an initial imperfection w_o (fig. 3.3). Because of the radial vertical stresses σ_v , this strip resists an axial load N and must shorten. The contributions to the shortening of the chord are respectively due to the elastic compressive strain, on the one hand, and to the non-linear geometric behaviour (bow effect), on the other hand; it is thus given as follows :

$$\delta = Nd / Et_w + 0.5 \int_o^d (\frac{\partial w}{\partial x})^2 \, dx \qquad (3.12.)$$

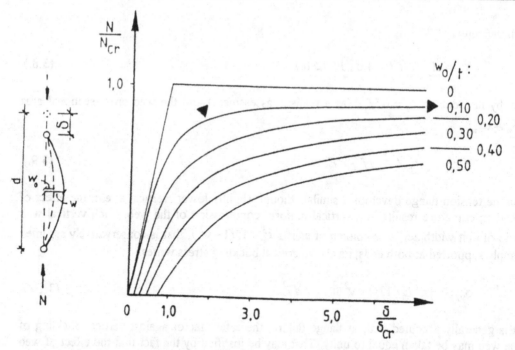

Figure 3.3 - Non-linear web strip behaviour

It is well-known that the total deflection amounts $w = w_o + w_{add}$ and writes :

$$w(x) = [w_o / (1 - N / N_E)] \sin(\pi x / d) \qquad (3.13.)$$

where N_E is the Euler column buckling load. Thus equation (3.12.) writes :

$$\delta = Nd / Et_w + (\pi^2 / 4d) [w_o / (1 - N / N_E)]^2 \qquad (3.14.)$$

The plot of δ against N/N_E (fig. 3.3) shows that when assuming $w_o \approx 0,1\,t_w$ to $0.2\,t_w$ - what is a realistic magnitude for the initial imperfection of a compressed strut -, the behaviour is quasi linear up to $N/N_E = 0.65$; it may reasonably be accepted that the vertical buckling of the flange into the web is prevented as far as the response of the web to vertical load remains quasi linear, i.e. :

$$N / N_E \equiv \sigma_v / \sigma_{v,cr} < 0.65 \qquad (3.15.)$$

This condition leads to a formula which is very close to equation (3.11.).

In Eurocode 3, the limit d/t_w ratio is written as follows :

$$d / t_w \leq k(E / f_{yf})\sqrt{A_w / A_{fc}} \qquad\qquad (3.16.)$$

where A_{fc} is the cross-sectional area of the compression flange and k, a factor the value of which depends on the class of section :

$k = 0.3$ for sections of Class 1
$k = 0.4$ for sections of Class 2
$k = 0.55$ for sections of Class 3 and 4

Factor $k = 0.55$ is the one obtained from both above reasonings, i.e. where the ultimate load in the flange is governed by the onset of yielding. For Classes 1 and 2, it is known that, due to plastic redistribution, the strain in the flange will exceed the yield strain. Basing the reasoning on the strain concept rather than on the stress concept would lead to more severe limits; the fact is indeed reflected by the above values of k specified by Eurocode 3.

The limit given by equation (3.16.) would be of course unduly severe, should the actual design stress in the flange be kept well below the yield stress of the flange material. That is the case in the vicinity of the end supports, where shear is dominant while bending is vanishing. Therefore it is allowed to substitute the design maximum stress in the flange for the yield stress f_{yf}.

For both previous reasonings, it has been assumed that there are no transverse stiffeners which could locally support the flange and possibly result in larger slenderness limit values. Such a beneficial effect is generally disregarded; it would anyway be small for usual spacings of transverse stiffeners.

When the girder is *curved in elevation*, with the compression flange on the concave side, the limit slenderness may be amended as follows :

$$d / t_w = k(E / f_{yf})\sqrt{A_w / A_{f.c}} / \sqrt{1 + dE / (3 r f_{yf})} \qquad\qquad (3.17.)$$

where r is the radius of initial in elevation curvature of the compression flange.

3.2. Resistance to transverse concentrated loads

In some circumstances, the web may be subject to *transverse concentrated loads*. Such loads exist necessarily at the supports; they may also occur in any section of the girder when special erection procedures are used (for instance, erection by launching).

When the concentrated load is applied in a specified section, a transverse stiffener may help the web in resisting this load. When the concentrated load is moving with respect to the girder, or conversely when the girder is moving with respect to the location of a concentrated load, then the web has to resist by its own.

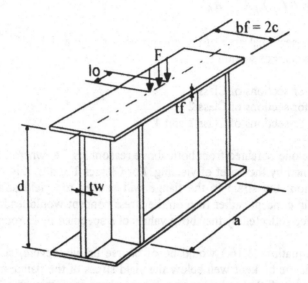

Figure 3.4 - Resistance to concentrated loads

Present section is concerned with the resistance of an unstiffened web to transverse concentrated forces. Three different modes of failure can be identified :

a) the *crushing* of the web in the close vicinity of the patch loaded flange, with plastic deformation of the flange;

b) the *crippling* of the flange, i.e. web buckling in the web zone located close to the patch loaded flange, with plastic deformation of the flange;

c) the web *buckling* over its whole depth.

When concentrated loads are applied, two cases may occur : either the loads are applied on one flange and are resisted by shear stresses in the web, or they are applied in both flanges in opposite directions so that they are transferred directly into the web. Here, only the first case (fig. 3.4) is worthwhile being considered, so that only failure modes (a) and (b) shall be investigated; in general, both ultimate loads are computed and the lowest one is taken as the resistance of the web to concentrated loads.

Generally transverse concentrated loads act coincidently with longitudinal bending. There is thus a need for investigating the possible interaction between transverse loading and longitudinal bending.

Crushing is the failure mode for a rather stocky web; the latter will indeed yield before it does buckle. The relevant failure load is governed by yielding over a length that is the sum of a *length of stiff bearing s_s* and a *distribution length s_y* (fig. 3.5).

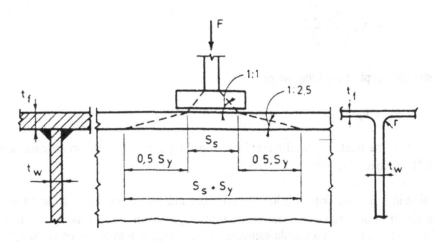

Figure 3.5 - Dispersion of a concentrated load

When computing s_s, a dispersion of load at a slope $1 : 1$ is assumed through solid steel material, provided it be properly fixed in place (no dispersion shall be taken through loose packs). Length s_y represents the plastic dispersion provided by the flange stiffness (refer to plastic mechanism in the flange). As an approximate the plastic dispersion may be assumed at a slope 1:2,5. There is some evidence that the flange flexural stiffness which can be mobilised corresponds to the part that is fully effective; therefore the restriction of b_f not larger than 25 times the flange thickness t_f (this value corresponds roughly to the limit b/t ratio for a flange - of a plate girder - in compression located at the interface of Class 2 and 3 Section). Above expression for s_y is used when the load is applied within the span of the girder; should the load be acting at the end member, s_y would be halved.

Much more determinative is the *crippling* load for girders with a slender web. The basic idea is to derive the general format for the ultimate patch load on base of a Von Karman - like approach, i.e. :

$$F_u = \sqrt{F_y F_{cr}} \leq F_y \qquad (3.18.)$$

where F_y and F_{cr} are the values of respectively the yield load and the elastic critical load of a web subject to a transverse point load resisted by shear stresses in the web.

Referring to a mechanism model with four plastic hinges for the case of a stocky girder subject to a concentrated load of short stiff bearing length s_s, one can derive :

$$F_y = l_{eff} t_w f_{yw} = (s_s + s_y) t_w f_{yw} \qquad (3.19.)$$

where :

$$s_y = 2t_f \sqrt{\frac{b_f}{t_w} \frac{f_{yf}}{f_{yw}}}$$

(3.20.)

Elastic critical plate buckling writes :

$$F_{cr} = k_F \, Et_w^3 / d$$

(3.21.)

where k_F is the buckling coefficient for the loading concerned; this coefficient amounts ..... for a "knife load" ($s_s = 0$).

Considering that the crippling load is no more influenced by the web depth when d / t_w exceeds 60 to 80 - this evidence results from experiments - and making due allowance for (3.20.) and (3.21.) with a simple estimate of l_{eff} ($l_{eff} = 6$ to $8t_f$), equation (3.18.) writes according to the general following format :

$$F_u = 0.5 \, t_w^2 \sqrt{Ef_{yw}} \sqrt{(t_f / t_w)}$$

(3.22.)

Several studies have been conducted in order to assess the influence of the length of stiff bearing. Eurocode 3 has suggested accordingly :

$$F_u = 0.5 \, t_w^2 \sqrt{Ef_{yw}} \left[\sqrt{t_f / t_w} + 3(t_w / t_f)(s_s / d) \right] / \gamma_{M1}$$

(3.23.)

where γ_{M1} is a partial safety factor and the ratio (s_s / d) cannot be larger than 0.2. Recent investigations conducted by *Duchene* by means of numerical simulations have demonstrated that above formula should be more appropriately written with a view to offer a better agreement with the results of the aforementioned numerical simulations; it was found to be :

$$F_u = 0.56 t_w^2 \sqrt{Ef_{yw}} \sqrt{t_f / t_w} (1 + 0.35 s_s / d) / \gamma_{M1}$$

(3.24.)

with the restrictions :

$$c \leq 12.5 \, t_f \qquad \text{otherwise } c \text{ shall be taken equal to } 12.5 \, t_f$$
$$s_s / d \leq 0.7$$
$$d / t_w \geq 80$$

Presently this expression is restricted to steel grades $f_y \leq 460 \, MPa$ because relevant information is still missing for higher steel grades.

Another promising formula has been obtained by Johanson and Lagerqvist. It is not derived, as the previous ones, from a plastic hinge mechanism but is directly found on an analytical expression, where the slenderness $\overline{\lambda} = \sqrt{F_y / F_{cr}}$ is the sole parameter. This alternative formula writes :

$$F_u = \chi F_y \tag{3.25.}$$

with :

$$\chi = 0.05 + (0.22 / \overline{\lambda}) + (0.21 / \overline{\lambda}^2) \tag{3.26.}$$

When, in addition to patch load F, the section is subject to a bending moment M, the following *interaction criterion*, suggested on base of test results, should be satisfied :

$$(F / F_u) + (M / M_u) \le 1.5 \tag{3.27.}$$

where F_u and M_u are the individual ultimate design resistances to patch load and bending respectively.

3.3. Resistance to direct stresses

A plate girder with a slender web cannot exhibit its full plastic bending resistance; indeed, the web slenderness is such that web buckling due to longitudinal direct stresses will generally occur well before complete plastic redistribution has taken place. Even the elastic yield resistance may not be reached because web buckling develops usually in the elastic range.

However there is a stress shedding from the web to the flanges in the postbuckling range, i.e. once the web has buckled. That results in a bending capacity higher than $\sigma_{cr} W_{el}$ where σ_{cr} is the elastic critical web buckling stress and W_{el} the relevant elastic section modulus (f_{yf} may be substituted for σ_{cr} when the tensile flange is governing). Of course a lower bound to the bending resistance can be assessed by fully neglecting the contribution of the web to the section modulus and assuming that the yield stress is reached in the determinative flange.

To make appropriate allowance for the loss in strength due to the effects of local buckling in the web, the compression flange is assumed to be proportioned so that to be fully effective, two approaches can be contemplated :
a) the use of the *effective stress concept*;
b) the use of the *effective width concept* applied to compression elements.

3.3.1. Effective stress concept

When use is made of the *effective stress concept*, calculations are conducted with the properties of the gross section, assuming a full efficiency of the plate components. Allowance is made for web buckling by limiting the design direct compressive stress to the relevant critical plate buckling stress. Account is taken for postbuckling strength by transferring the load from the web - which buckles - to the flange - which has presumably not yet yielded -. This load shedding is allowed for provided it does not exceed a certain proportion; it depends also on the boundary conditions and of course equilibrium of the section must still be fulfilled after load shedding has taken place. The concept of effective stress has mainly be promoted in U.K. On the one hand, this concept could be sometimes misunderstood and therefore misused in practice. On the other hand, the concept of effective width has been adopted in most national codes and more recently in Eurocode 3; the concept of effective stress has therefore lost much of its attractiveness and is not worthwhile being examined further.

3.3.2. Effective width concept

The *effective width concept* has been first introduced for a simply supported plate subject to uniform uniaxial compression. It has been generalized afterwards to other types of loading.

3.3.2.1. Perfect plate in uniaxial uniform compression

Let us thus first consider a perfectly flat rectangular simply supported plate element subject to uniform compression in one direction. The elastic plate critical buckling stress is :

$$\sigma_{cr} = k_\sigma \left[\pi^2 E / 12(1-v^2)\right](t/b)^2 \qquad (3.28.)$$

where b/t is the width-to-thickness ratio and k_σ the buckling coefficient (see Lecture 1) :

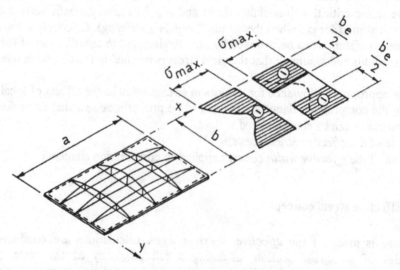

Figure 3.6 - Concept of effective width

Once this stress is reached in the element, the latter may buckle. The longitudinal bow effect, the magnitude of which varies across the plate width, results in a non uniform stress distribution (fig. 3.6). All the deflected longitudinal fibres have indeed no more the same stiffness in the postbuckling range. It is widely accepted that the ultimate load is close to the loading level at which the membrane stress at both longitudinal edges reaches the material yield stress. Therefore this simple criterion can be adopted as a rather realistic definition of the ultimate limit state.

The resultant of the stress distribution allows to compute the *average* ultimate compressive stress $\overline{\sigma}_u$ according as :

$$N_u \equiv 2t \int_0^{b/2} \sigma\,(y)\,dy = b.t.\overline{\sigma}_u \tag{3.29.}$$

The computation of the integral involved in (3.29.) requires that the shape of the stress distribution be known. In order to obviate this difficulty, the concept of effective width b_e, smaller than actual width b, is introduced. When subject to a constant stress $\sigma_{max} = f_y$, the effective width would transmit the same stress resultant N_u. Thus :

$$N_u = b_e.t.f_y \tag{3.30.}$$

Combining (3.29.) and (3.30.) results in :

$$b_e\,/\,b = \overline{\sigma}_u\,/\,f_y \tag{3.31.}$$

On the other hand, *von Karman* has postulated that the effective width b_e is the width of a fictitious plate which has same thickness and aspect ratio as the real plate and which would buckle for an elastic critical plate buckling stress equal to $\sigma_{max} = f_y$. Accordingly, one can write :

$$\sigma_{cr}(b_e) \equiv k_\sigma\left[\pi^2 E\,/\,12(1-v^2\,)\right](t\,/\,b_e\,)^2 = f_y \tag{3.32.}$$

Combining (3.28.) and (3.32.) results in :

$$b_e\,/\,b = \sqrt{\sigma_{cr}\,/\,f_y} \le 1 \tag{3.33.}$$

or, account taken of (3.31.) :

$$\overline{\sigma}_u = \sqrt{\sigma_{cr}f_y} \tag{3.34.}$$

According to *von Karman*, the average ultimate stress would thus be given as the geometric mean of both the critical plate buckling stress and material yield stress.

As the *normalized plate slenderness* writes (see Lecture 2) :

$$\bar{\lambda}_p = \sqrt{f_y / \sigma_{cr}} \qquad\qquad (3.35.)$$

while the *normalized ultimate load* is given as :

$$\bar{N}_p \equiv N_u / N_{pl} \equiv b_e / b \leq 1 \qquad\qquad (3.36.)$$

it comes :

$$\bar{N}_p = 1 / \bar{\lambda}_p \leq 1 \qquad\qquad (3.37.)$$

The latter expression is known as *von Karman formula* written in terms of normalized coordinates; it is to the perfect plate what *Euler formula* $\bar{N} = 1/\bar{\lambda}^2$ is to the perfect column. The range comprised between both curves reflects the increase in ultimate load which results from the postbuckling behaviour of a perfect plate.

Should the design maximum compressive stress σ_{max} be significantly lower than f_y, then a safe approach would consist in substituting σ_{max} for f_y in the plate slenderness $\bar{\lambda}_p$.

3.3.2.2. Effect of plate imperfections

Geometric imperfections - such as mainly an initial out-of-flatness - and structural imperfections - amongst which residual stresses are the most determinative - cause non-linearities and precipitate first yielding.

In order to account for such effects, above von Karman formula must be amended. Based on a series of tests on cold-formed members conducted in the fourties, *Winter* suggested the following expression :

$$\bar{N}_p \equiv b_e / b = \left[1 - 0{,}22 / \bar{\lambda}_p \right] / \bar{\lambda}_p \leq 1 \qquad\qquad (3.38.)$$

In the late sixties, *Faulkner* proposed a similar expression derived from tests on stiffened plates used in naval architecture, which gives only slightly different results compared to *von Karman* formula :

$$\bar{N}_p = 1{,}05 \left[1 - 0{,}26 / \bar{\lambda}_p \right] / \bar{\lambda}_p \leq 1 \qquad\qquad (3.39.)$$

The range of application for *Winter formula*, to which it is now widely referred, is $\overline{\lambda}_p > 0.673$; indeed, for $\overline{\lambda}_p < 0.673$, yielding is governing and $\overline{N}_p = 1$.

3.3.2.3. Generalization to a linear distribution of direct stresses

There is some evidence that Winter formula is likely to be applied to plates subject to a linear distribution of direct stresses, provided the stress ratio $\psi = \sigma_2 / \sigma_1$ (σ_1 being the largest compressive stress and σ_2 the lowest compressive or maximum tensile stress) be somewhere accounted for.

When there is a sign reversal in the direct stress distribution over the width b, only the portion b_c, subject to compressive stresses, has to be considered whereas the portion $(b-b_c)$, subject to tensile stresses, is fully effective. Therefore expression (3.38.) may be used provided that b_c be substituted for b :

$$b_e / b_c = \left[1 - 0.22 / \overline{\lambda}_p\right] / \overline{\lambda}_p \leq 1 \tag{3.40.}$$

An alternative to (3.40.) is given in *Publication n° 44 of E.C.C.S.*, it writes :

$$b_e / b_c = \left[1 - 0.05(3 + \psi) / \overline{\lambda}_p\right] / \overline{\lambda}_p \leq 1 \tag{3.41.}$$

In (3.40.) the influence of the stress ratio is only reflected implicitly through $\overline{\lambda}_p$; in (3.41.), there is an additional explicit effect through factor $0.05(3+\psi)$.

Formula (3.41.) accounts for a less severe influence of imperfections in bending than in pure compression; that is generally accepted indeed. Despite this fact, Eurocode 3 has adopted Winter formula with a simplified format : factor 0.2 is substituted for 0.22 one in (3.38.).

Once the effective width b_e of the compression zone b_c is determined, it must be allocated by appropriate portions b_{e1} and b_{e2} which are however limited to $b_{e1} \nless 0.4b_e$ and $b_{e2} \ngtr 0.6b_e$. These portions are respectively adjacent to the most and to the least compressed fibres of the compression zone, when the plate is supported on both longitudinal edges (*internal compression elements*). For *outstand compression elements*, which have one simply supported longitudinal edge and a free one, the whole effective width of the compression zone is adjacent either to the least compressed fibre or to the most compressed fibre, according as the stress gradient.

The portions b_{e1} and b_{e2} are computed as :

$$\begin{aligned}
b_{e1} &= 2\rho b / (5 - \psi) \leq 0.4b_e \\
b_{e2} &= (3 - \psi)\rho b / (5 - \psi) \leq 0.6b_e
\end{aligned} \tag{3.42.a,b}$$

where $\rho = b_e / b_c$.

How to apply extensively the concept of effective width when assessing the resistance of plate and box-girders is explained in Lecture 5.

3.4. Resistance to shear stresses

The resistance of the web to shear may be governed either by yielding or by web buckling. The distinction between both collapse modes is based on the depth-to-thickness ratio, d/t_w, i.e. on the plate slenderness ratio $\overline{\lambda}_p$. For shear, the latter writes :

$$\overline{\lambda}_p = \sqrt{\tau_y / \tau_{cr}} \le 1 \tag{3.43.}$$

where the *yield shear stress* amounts $f_{yw} / \sqrt{3}$.

It is usually assumed that the range of $\overline{\lambda}_p$ for a full efficiency in shear is a bit larger than for direct stresses; it is taken equal to 0.8 in Eurocode 3. It may thus be concluded as follows for what regards the limit depth-to-thickness : there is no effect of plate buckling and the plate shear resistance will be the yield shear load when $\overline{\lambda}_p \le 0.80$. Account taken that the *plate critical shear buckling stress* of a web is :

$$\tau_{cr} = k_\tau \left[\pi^2 E / 12(1 - v^2) \right] (t_w / d)^2 \tag{3.44.}$$

the condition $\overline{\lambda}_p \le 0.80$, leads to :

$$d / t_w \le 30 \, \varepsilon \sqrt{k_\tau} \tag{3.45.}$$

For a *simply supported unstiffened web*, the aspect ratio is such that k_τ approaches 5.34; above limit writes simply :

$$d / t_w \le 69 \varepsilon \tag{3.46.}$$

This limit is generally exceeded for webs of welded plate -and box-girders , so that the buckling resistance of the web is then mostly determinative.

How to evaluate the shear buckling resistance of webs, account taken of postbuckling strength reserve, is reviewed in Lecture 4.

LECTURE 4

ULTIMATE SHEAR RESISTANCE OF WEBS

1. INTRODUCTION

The main role of the web is to resist shear. When stocky, the web can develop its plastic shear resistance. When very slender, it buckles before yielding but is then prone to exhibit a large postbuckling strength, with the result that the shear capacity may exceed appreciably the elastic critical shear buckling load. Should the shear capacity of the web exceed notably the design shear load, then the web material would be able to resist coincident direct and shear stresses and thus contribute the bending resistance of the section.

The critical shear buckling stress of a thin web having the dimension *(a x d)* writes :

$$\tau_{cr} \equiv k_\tau \sigma_E = k_\tau \left[\pi^2 E / 12(1 - v^2) \right] (t_w / d)^2 \qquad (4.1.)$$

d is the web depth, t_w is the web thickness and a is the spacing of the transverse stiffeners. The buckling coefficient k_τ depends on the aspect ratio $\alpha = a / d$, on the support conditions and possibly on the longitudinal web stiffener properties. Usually the web of a plate girder is conservatively assumed simply supported along its four edges; accordingly k_τ writes for a non longitudinally stiffened web :

$$k_\tau = 4 + 5.34 / \alpha^2 \qquad (\alpha \le 1)$$
$$= 5.34 + 4 / \alpha^2 \qquad (\alpha \ge 1)$$

When the length of the plate is large compared to the width *(α>>1)*, the buckling coefficient approaches 5.34.

It has been established in Lecture 3 that shear buckling resistance will govern the web design to shear when :

$$d / t_w \ge 30 \varepsilon \sqrt{k_\tau} \qquad (4.2.)$$

where $\varepsilon = \sqrt{235 / f_{yw}}$.

For plate slenderness d/t_w lower than the limiting value given by (4.2.), yield would be governing. However the material behaves elastically up to a limit of proportionality only; in addition account must be taken of the detrimental influence of residual stresses, which initiate prematurous first yielding. For an *imperfect plate*, there is thus a range where plate buckling is interacting with yielding. Accordingly, as soon as τ_{cr} exceeds a proportion β of the shear yield stress, a reduced critical buckling stress τ_{cr}^{red} must be substituted for τ_{cr}. Many expressions have been suggested in this respect; the one recommended by Eurocode 3 uses $\beta = 0.67$ and writes, in normalized coordinates :

$$\tau_{cr}^{red} / \tau_{yw} = \left[1 - 0.8(\overline{\lambda}_w - 0.8)\right] \not> 1$$
$$\not> 1 / \overline{\lambda}_w^2 \tag{4.3.}$$

with the normalized plate slenderness :

$$\overline{\lambda}_w = \sqrt{\tau_{yw} / \tau_{cr}} \tag{4.4.}$$

2. PHYSICAL BACKGROUND

How a slender web can exhibit a postbuckling shear strength is explained here below. Let us refer to a rectangular slender web, which is assumed to be perfectly flat and without residual stresses and is subject to in-plane pure shear. These assumptions aim at facilitating the reasoning and at understanding better the complete plate behaviour up to collapse. They are not at all compulsory; indeed usual geometrical and structural imperfections remain mostly without significant effect on the ultimate shear strength.

When shear load grows up, the web buckles elastically when the shear stress reaches the critical shear buckling stress τ_{cr}. Till this stage, the web resists by *beam action* strength, similarly to the web of any rolled section. Pure shear in the web is equivalent to equal but opposite principal direct stresses σ_1 and σ_2, acting at 45° with respect to shear; shear buckling may thus be understood as produced by principal compressive stresses $\sigma_2 = -\tau_{cr}$ (fig. 4.1.a.).

Once the web has buckled, it is no more able to resist any additional direct stress in the direction of σ_2, while there is still tensile strength reserve in the tensile direction. Henceforth the web is able to act at the manner of the tensile diagonal of a Pratt truss; the chords and the posts of this truss are constituted by the flanges and the transverse stiffeners respectively. Otherwise speaking, the shear load can increase beyond the critical shear buckling resistance of the web, provided there is, subsequently to buckling, a change in the stress distribution. An appreciable postbuckling strength may be mobilized because of the diagonal tensile stresses that develop. This second stage corresponds to the *tension-field action* strength (fig. 4.1.b.).

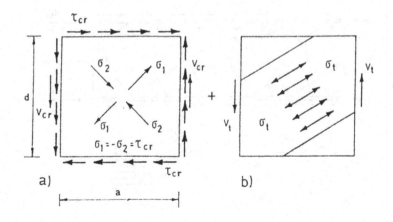

Figure 4.1 - Contributions to ultime shear resistance

The principal tensile stresses can grow up to the material yield stress. That results in principal tensile and compressive stresses of different magnitude, which are no more sloped at 45° on the web edges (fig. 4.2). The web edges experience not only shear stresses but also direct stresses; the latter must be resisted by the edge members (flanges and transverse stiffeners). If the latter are *infinitely rigid*, the postcritical strength may develop uniformly over the whole area of the web. When, in contrast, the edge members are very flexible, there are less facilities for anchoring the direct stresses; only a reduced postbuckling strength reserve can be mobilised. In the extreme case of a *very slender* web ($t_w / d \rightarrow 0$), the elastic critical shear stress τ_{cr} is so small that it may be neglected. Accordingly the principal compressive stress σ_2 is zero. Shear load is then sustained by the sole postbuckling behaviour, i.e. by tensile principal stresses in the web plate. This mode of resistance is termed *tension field* and the *tension field theories* are the methods used to describe the corresponding behaviour.

The ultimate shear resistance of a plate girder appears as the superimposition of the critical shear buckling resistance and the tension field contribution. It requires consequently the knowledge of :

a) the *magnitude of tensile and compressive principal stresses*;

b) the *direction of the principal stresses*;

c) the *contribution of edge members* to the postbuckling strength.

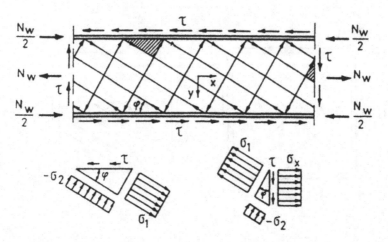

Figure 4.2 - Stress state in a web plate

3. TENSION FIELD APPROACHES

3.1. The pioneers

Rode (1916) has been the first pioneer of the tension field concept. His approach is rather rough; it is assumed that the tension field has a width of 50 times the web thickness t_w.

Wagner (1929) has developed the *ideal tension field theory*, which approaches the plate in shear as a system of ties covering the *whole* area of the web. This theory is quite appropriate for *very thin plates* ($d/t_w = 600$ up to 1000) connected to *very rigid* edge members. Prior to plate buckling, the shear stress distribution is pure shear : it is represented by a Mohr circle of radius $\tau \leq \tau_{cr}$ centered at the origin. Because of the extreme thinness, σ_2 is negligible; once the web has buckled, the representative Mohr circle shifts to the right, and is tangent to the τ axis (fig. 4.3.a.). At collapse, $\tau = \tau_u$ and the principal tensile stress amounts :

$$\sigma_1 = 2\tau_u \tag{4.5.}$$

and develops over the whole web. The Wagner shear resistance writes :

$$V_u = \sigma_1 t_w g \sin\theta \tag{4.6.}$$

where θ and $g = d\cos\theta$ are respectively the slope and the width of the tension field. The maximum value of V_u corresponds to :

i)   the slope θ for which the first derivative $\partial V_u / \partial\theta$ vanishes, and

ii)   tensile yielding ($\sigma_1 = f_{yw}$) in the tension field.

Then the principal tensile stresses are sloped 45° and the ultimate shear amounts :

$$V_{u,max} = 0.5\ dt_w\ f_{yw} \qquad\qquad\qquad (4.7.)$$

i.e. 13 % less than the full yield shear resistance.

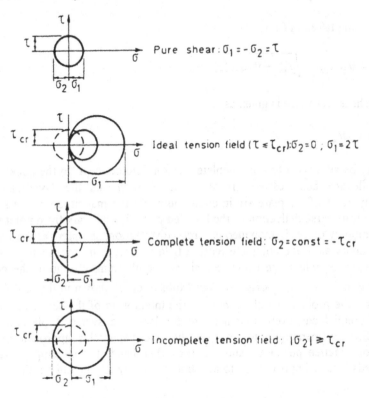

Figure 4.3 - Mohr circle for several tension field models

Above theory had to be improved. Indeed steel plates used in civil engineering structures have usually i) edge members of finite bending rigidity, and ii) a non zero critical shear buckling resistance.

In the *complete tension field theory*, the edge members are still assumed fully rigid but due allowance is made for a non zero critical shear resistance; therefore $|\sigma_2| = \tau_{cr} \neq 0$. The representative Mohr circle (fig. 4.3.b.) shows that :

$$\sigma_1 = 2\tau_u - \sigma_2 \qquad\qquad\qquad (4.8.)$$

At collapse, the value of σ_l is governed by the *von Mises yield criterion* :

$$\sqrt{\sigma_1^2 - \sigma_1\sigma_2 + \sigma_2^2} = f_{yw} \tag{4.9.}$$

wherefrom :

$$\sigma_1 = 0.5 \, \sigma_2 + \sqrt{f_{yw}^2 - 0.75 \, \sigma_2^2} \tag{4.10.}$$

and, account being taken of (4.8.) :

$$2\tau_u = 0.5 \, \tau_{cr} + \sqrt{f_{yw}^2 - 0.75 \, \tau_{cr}^2} \tag{4.11.}$$

The ultimate shear resistance is given as :

$$V_u = \tau_u dt_w \tag{4.12.}$$

Criticisms may be addressed to the complete tension field concept. In the buckled plate, the principal tensile and compressive stresses must be such that the deviation forces they produce at any point of the plate are in equilibrium. So the magnitude of the stresses shall depend on the transverse deflection of the buckled plate; it shall vary from point to point, so that the assumption of an homogeneous stress distribution, adopted as well in the ideal tension field theory as in the complete tension field theory, is no more valid. The principal compressive stresses, which are liable to maintain equilibrium, change in the postbuckling domain, so that $|\sigma_2| > \tau_{cr}$; the corresponding Mohr circle is shown in figure 4.4.c. The sole way for solving the problem would consist in the integration of the set of the two coupled 4th order differential equations, that account for large transverse displacements and the stabilizing effect of membrane forces. On the one hand, this approach is not at all appropriate for practice purposes, and, on the other hand, the uncomplete tension field model still needs further improvement to account explicitly for the finite rigidity of the edge members.

4. TENSION BAND APPROACHES

In the field of civil engineering, the assumption that the stiffness of the edge members is very large compared to the one of the web is not realistic. It was explained above that in the postbuckling range, the web edge members of a plate girder have to resist shear and direct stresses. These edge members (flanges, transverse stiffeners) are consequently prone to bend and to move in the web plane. The in-plane displacements of the edge members result first, in a drop in magnitude of the membrane stresses - compared to the case of infinitely rigid edge members - and second, in a change in direction of the principal stresses. When adjacent web panels, the direct stresses which develop in one panel can anchor efficiently because of the presence of adjacent panels; then in-plane displacements of transverse stiffeners are *virtually* prohibited and their effect can be disregarded. That is especially the situation of intermediate transverse web stiffeners; in contrast, end stiffeners do usually not benefit from such a favourable situation.

A further improvement of the tension field theories should be to account properly for the ability to in-plane displacements of the edge members.

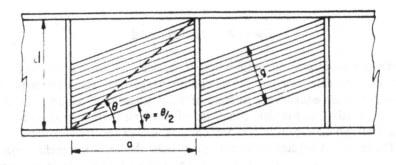

Figure 4.4 - Basler model

Basler has developed a tension field model, where the rigidity of both flanges is fully neglected; the tension field anchors thus on the sole transverse stiffeners if they are (fig. 4.4). According to this approach, it is assumed that the plate critical shear resistance :

$$V_{cr} = dt_w \tau_{cr}$$ (4.13.)

remains mobilised by the web plate in the postbuckling range, irrespective of the tension field action. The ultimate shear load will be reached when the web material yields under coincident pure shear stresses and direct stresses developed by the tension field action. In addition, the frame composed of flanges and stiffeners contributes partially the strength capacity by *frame action* (*Vierendeel action*). Basler's theory leads to ultimate shear loads, which are only found in good agreement with experimental results when the flange stiffness is large compared to the web stiffness.

Tests conducted by *Rockey and Skaloud* on plate girders of same geometry as those tested by *Basler* but with flanges of different proportions, have shown that the bending stiffness of the flanges is likely to influence appreciably the shear capacity. Many researchers have tried to improve Basler's model accordingly. It is not the place here to review all the work done in this respect; let us just point out the *Cardiff model*, which is probably the best known. It is still assumed that the critical shear strength V_{cr} is still mobilised till the ultimate load is

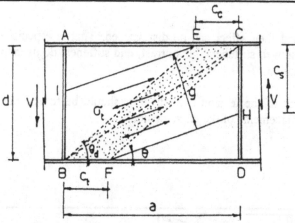

Figure 4.5 - Cardiff model

reached and super-imposes to the tension field resistance V_t. Thus the tension field anchors not only on the transverse stiffeners but also on the flanges. To which extent the anchor to flanges is concerned depends on the stiffness of these flanges with regard to bending in the web plane. The *tensile band* is thus composed of a central portion, that anchors onto the stiffeners, and of two additional lateral portions that anchor respectively onto both flanges (fig. 4.5.). The larger the bending stiffness of the flanges, the larger the anchor lengths c_c and c_t on the flanges. The values of c_c and c_t are computed based on the consideration of a combined failure mechanism with plastic hinges in the flanges and a yielded band in the web (fig. 4.6. and 4.7.). Therefore, the effect of the flange stiffness is reflected through the plastic bending resistance of the flanges; because this resistance is influenced by possible axial forces in the flanges, the anchor lengths on the flanges shall be reduced according as the magnitude of direct stresses in the flanges.

The contribution of the tension band to the ultimate shear load is the vertical component of the resultant force in this band, i.e.

$$V_t = g t_w \sigma_t \sin\theta \tag{4.14.}$$

where σ_t is the magnitude - at collapse - of the presumed uniform tensile stress distribution across this band of slope θ.

The width g of the tensile band can be determined as a function of : i) the dimensions a and d of the web panel; ii) the inclination θ of this band; iii) the anchor lengths c_c and c_t on the flanges :

$$g = d\cos\theta - (a - c_c - c_t)\sin\theta \tag{4.15.}$$

Plastic hinges in the upper and lower flanges are expected to occur in sections E and F (fig. 4.5 and 4.7), where flange shear force vanishes, and in sections C and B, where the flanges are considered as fully clamped.

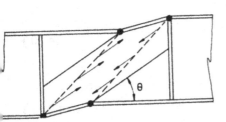
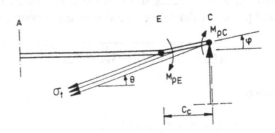

Figure 4.6 - Combined mechanism   Figure 4.7 - Plastic mechanism in the flange

Applying the principle of virtual work to the corresponding plastic mechanism in the upper flange gives (fig. 4.7) :

$$0.5 \, \sigma_t t_w c_c^2 \, \phi \, sin^2 \, \theta = (M_{pE}^* + M_{pC}^*) \, \phi \qquad (4.16.)$$

where M_p^* is the plastic bending resistance of the flanges, with account taken of the possible axial load N_f in this flange :

$$M_p^* = M_{pf} \left[1 - (N_f / N_{pf})^2 \right] \qquad (4.17.)$$

Length c_c is drawn from (4.16.); so is also c_t :

$$c_c = (2 / sin\theta) \sqrt{(M_{pE}^* + M_{pC}^*) / 2\sigma_t t_w} \ngtr a \qquad (4.18.a.)$$

$$c_t = (2 / sin\theta) \sqrt{(M_{pB}^* + M_{pF}^*) / 2\sigma_t t_w} \ngtr a \qquad (4.18.b.)$$

Usually, M_{pB}^* and M_{pF}^*, on the one hand, and M_{pE}^* and M_{pC}^*, on the other hand, are not much different so that the anchor lengths write more simply in the general following format :

$$c = 2 / sin\theta \sqrt{M_p^* / \sigma_t t_w} \qquad (4.19.)$$

where M_p^* is an *average* reduced bending resistance of the relevant flange.

The stress σ_t in the tension band at collapse is derived from the *von Mises yield criterion* written with reference to the stress state related to directions u and v, respectively parallel and perpendicular to the direction of the tension band :

$$\sqrt{\sigma_u^2 + \sigma_v^2 - \sigma_u\sigma_v + 3\tau_{uv}^2} = f_{yw} \tag{4.20.}$$

where :

$$\sigma_u = \sigma_t + \tau_{cr}\,sin\,2\theta \tag{4.21.a.}$$
$$\sigma_v = -\tau_{cr}\,sin\,2\theta \tag{4.21.b.}$$
$$\tau = \tau_{cr}\,cos\,2\theta \tag{4.21.c.}$$

Introducing (4.21.) in (4.20.) gives :

$$\sigma_t = \sqrt{f_{yw}^2 - \tau_{cr}^2\left[3 - 2,25\,sin^2\,2\theta\right]} - 1,5\tau_{cr}\,sin\,2\theta \tag{4.22.}$$

where τ_{cr}^{red} , given by (4.3.) has to be possibly substituted for τ_{cr} when necessary.

At this stage, the sole remaining unknown is the slope θ of the tensile band. Its value should maximize V_t because of the static theorem of plastic design (indeed a state of stress has been defined which is in equilibrium with the shear load and complies with the yield condition). Its search ought thus to proceed by trial and error; however numerous computations have shown that the part of the curve $V_t = f(\theta)$ is very flat in the vicinity of its maximum. Thus any error on the assessment of θ_{opt} will only generate a slight error on V_t but on the safe side. It is observed experimentally that θ_{opt} is always very close to 2/3 θ_d, where θ_d is the slope of the geometric diagonal of the web panel :

$$\theta_d = arctg(d\,/\,a) \tag{4.23.}$$

This value can thus be used as a conservative approximate. Alternatively iterations may be used to find the actual optimum value θ_{opt}.

Frame action is a possible third contribution V_f to the ultimate shear resistance. When the tensile diagonal of the pseudo-Pratt truss has exhausted its carrying capacity (because of yielding), the frame constituted by the flanges and the transverse stiffeners can still be expected to resist by *Vierendeel action*; it can thus experience some shear load till the onset of plastic hinges in this frame with the result of a plastic collapse mechanism. Frame action is usually disregarded because of the two main reasons : i) its contribution V_f is usually much smaller than both other ones V_{cr} and V_t, and ii) experiments and numerical analysis show that the real collapse of the surrounding frame - which implies onset of plastic hinges - develops after the maximum shear load has been overcome.

Based on calibration against test results, the design shear resistance shall be limited to :

$$V_u = (V_{cr} + 0.9\,V_t) / \gamma_{M1}$$ (4.24.)

where γ_{M1} is a partial safety factor, equal to 1.1 in accordance with Eurocode 3.

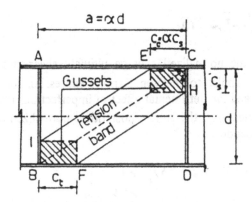

Figure 4.8 - Dubas' model

Dubas has developed an ultimate shear model, which it is termed *simple postcritical method* because of its simplicity. The basic idea regarding the two successive steps of the behaviour is still the same; the difference lies in the assessment of the postbuckling strength reserve. Here the tension band is oriented according as the geometric diagonal of the web panel (fig. 4.8); it is anchored on two rectangular gusset plates, which have the same aspect ratio as the web panel and have dimensions such that the critical shear stress of these gusset plates is equal to the material yield stress wherefrom :

$$c_s / d = \sqrt{\tau_{cr} / \tau_{yw}}$$ (4.25.)

where τ_{cr} is the critical shear buckling stress of the web panel. The additional shear contribution as a result of the tension field action is provided by the gusset plate; it is limited by shear yielding of this gusset plate. It writes :

$$V_t = c_s t_w (\tau_{yw} - \tau_{cr})$$ (4.26.a.)

or, account taken of (4.25.) :

$$V_t = d\, t_w \sqrt{\tau_{cr} / \tau_{yw}} (\tau_{yw} - \tau_{cr})$$ (4.26.b.)

The frame action is disregarded too so that the shear buckling resistance writes :

$$V_u = V_{cr} + V_t$$

$$= dt_w \sqrt{\tau_{cr}\tau_{yw}} \left[1 + \sqrt{\tau_{cr}/\tau_{yw}} - (\tau_{cr}/\tau_{yw}) \right] \tag{4.27.}$$

For large aspect ratios, the term between brackets is close to 1.0; then the design shear buckling resistance writes conservatively :

$$V_u = 0.9 \, dt_w \sqrt{\tau_{cr}\tau_{yw}} / \gamma_{M1} \tag{4.28.}$$

which corresponds to the normalized average ultimate shear stress $\tau_u / \tau_{yw} = 0.9 / \overline{\lambda}_w$. Equation (4.28.) looks similar to the von Karman expression for pure compression. To account for inelasticity and the detrimental effect of imperfections (in the intermediate range of web slenderness values $\overline{\lambda}_w$), the simple postcritical normalized *ultimate shear stress* is given as follows in Eurocode 3 :

$$\tau_u / \tau_{yw} = \left[1 - 0.625(\overline{\lambda}_w - 0.8) \right] \begin{array}{l} \not> 1 \\ \not> 0.9 / \overline{\lambda}_w \end{array} \tag{4.29.}$$

which is slightly different from the format (4.3.) used for elastic critical shear buckling stress.

5. EVIDENCE FOR THE NEED OF BOTH APPROACHES

All the test results dealing with plate girders subject to shear have been collected and compared with the theoretical ultimate shear forces, computed respectively in accordance with the Cardiff tension band model, on the one hand, and with the simple postcritical method, on the other hand. Plotting the values of the ratio V_{ex} / V_u against the values of aspect ratio $\alpha = a/d$ has shown that :

a) the tension field model is appropriate for transversely stiffened plate girders whose web panels aspect ratio is in the range 1 to 3; it is too conservative for large values of α and unsafe for values of α lower than 1;
b) the simple postcritical method is appropriate for large values of the aspect ratio (> 3) and unduly conservative for smaller values of α.

For large spacings of the transerse stiffeners ($\alpha > 3$), the tension band is found to develop with a slope higher than that computed based on the tension field approach and no more in relation with the aspect ratio.

It is also worth while investigating whether the different parameters are appropriately accounted for in both methods.

5.1. Comparison between test results and tension field approach

Values of V_{ex}/V_u have been plotted against web depth d, web thickness t_w, web slenderness $\overline{\lambda}_w$ (60-400) and material yield stress f_y (235-600 N/mm^2). It may be concluded that the average value and the standard deviation are approximately constant in the whole range of variation of the parameters.

Also the influence of relative flange stiffness on the ratio V_{ex}/V_{pl} has been studied. The relative flange stiffness parameter has been defined as $I_f.10^6/a^3 t_w$, where I_f is the moment of inertia of the flange. From this comparison, it results that the choice of a slope of the tension band which does not depend on the spacing of the transverse stiffeners is just an approximate which is justified because the influence is only slightly marked.

5.2. Comparison between test results and simple postcritical approach

The plot of the reduced shear capacity against the plate slenderness has shown that the model for τ_u, i.e. $\tau_u/\tau_{yw} = 0.9/\overline{\lambda}_p$ is conservative. The dependency of the experimental to theoretical ultimate shear ratio versus the aspect ratio shows a slight tendency to increase when α increases.

6. AMENDMENT WHEN BOX-GIRDERS

Because box-girders have usually very wide flanges, the question arises to determine to which extent these flanges contribute effectively to the anchor of the tension band.

Provisionally and for sake of safety, it is recommended to disregard any such anchor $(c_c = c_t = 0)$ and to adopt a slope θ of the tension bond equal to half the slope θ_d of the geometric diagonal $(\theta = \theta_d/2)$.

LECTURE 5

RESISTANCE TO DIRECT STRESSES

1. INTRODUCTION

The rules presented hereafter are dealing with the Ultimate Limit States verification of girders which experience compression forces, bending and shear. They have of course to account for plate buckling.

The effects of plate buckling are considered by means of the concept of effective width which must be determined for each individual compression plate element of the girder. It is explained in Lecture 2 that the effective width depends on the edge support conditions and on the gradient of direct stresses; in addition the following effects should be taken into account :

a) the interaction between possible shear lag and plate buckling;
b) the column-like behaviour of narrowly spaced panels loaded in compression, or of stiffened panels;
c) the interaction between coincident direct and shear stresses.

Item c) is the sole with which the plate girders are concerned in practice; in this respect, only non longitudinally stiffened plate girders are examined.

Items (a) to (c) are especially of concern in wide stiffened compression flanges (box girders).

2. CROSS-SECTIONAL PROPERTIES OF CLASS 4 SECTIONS

The cross-sectional properties of a Class 4 section are computed based on the *effective cross-section*; the latter involves the effective width of all the compression elements calculated in accordance with the rules given in Lecture 2.

To determine the effective width of the compression flange, the stress is computed based on the stress ratio ψ determined from the properties of the gross cross-section; indeed the web does not contribute very much the section moduli and therefore the stress distribution. It is however usual to assume that the yield stress will be reached at the ULS and to determine the effective width accordingly. To determine the effective width of the web, use shall be made of the stress ratio ψ obtained by considering the effective area of the compression flange(s) but the gross area of the web.

Generally the centroïdal axis of the effective cross-section has shifted by an amount e compared to the centroïdal axis of the gross cross-section. This must of course be accounted for when calculating the elastic bending properties of the effective cross-section (fig. 5.1). When the cross-section is subject to an axial force, this force is usually referred to the centroïd of the gross cross-section, as a result of the global analysis. Thus, bending moment ΔM :

$$\Delta M = N e_N \qquad (5.1.)$$

must be considered when the centroïd of the effective cross-section for uniform compression only has shifted by an amount e_N compared to the centroïd of the gross-section (fig. 5.2). Therefore it is not unusual that a section presumably subject to an axial force must be checked for combined compression and bending.

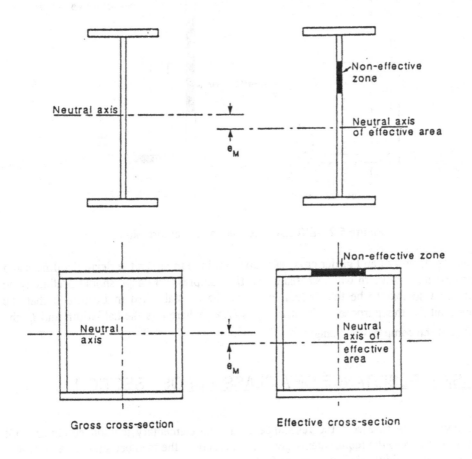

Figure 5.1 - Effective section : pure bending

For greater economy, the plate slenderness $\bar{\lambda}_p$ of any compression element may be determined by substituting the maximum compressive stress in that element for the relevant material yield stress f_y. Such an allowance is permitted provided that this stress be consistent with the effective section which is searched. This procedure is lengthy; it requires indeed an iterative trial and error process in which ψ is calculated again at each step - on base of the stresses computed in the effective cross-section obtained at the previous step -, till a complete consistency be reached.

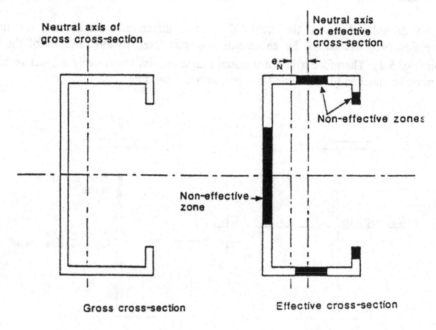

Figure 5.2 - Effective section : axial compression

Because the flanges of an I girder cross-section contribute the most the flexural stiffness and section moduli, it is worth while stressing that the compression flange should preferably be proportioned so that to be fully effective. It has been established in Lecture 2 that the relevant limit for this purpose is about $c / t_f \leq 12.5$, where c is the half-width and t_f the thickness of the compression flange.

3. DESIGN RESISTANCE OF A CLASS 4 CROSS-SECTION

Let us review how to assess the resistance to all the elementary loading conditions. Of course, in addition to the requirements given in this section, the member stability and, where necessary, frame stability shall also be verified.

3.1. Resistance to tensile force

Plate buckling is not of concern. The design tension resistance N_{tRd} of a Class 4 cross-section is the smaller of :

a) the *design plastic resistance of the gross cross-section* :

$$N_{pl,Rd} = A f_y / \gamma_{MO} \qquad\qquad (5.2.a.)$$

b) the *design ultimate resistance of the net cross-section* at bolt holes :

$$N_{u,Rd} = 0,9 A_{net} f_u / \gamma_{M2} \qquad\qquad (5.2.b.)$$

where A is the cross-sectional area of the gross section, A_{net} that of the net section and f_u the ultimate tensile strength of the material. Coefficients γ_{MO} and γ_{M2} are partial safety factors; according to Eurocode 3, they are respectively $\gamma_{MO} = 1.00$ and $\gamma_{M2} = 1.25$.

Most often a ductile behaviour is required; then N_{pl} must be less than N_u, wherefrom :

$$0,9 \, (A_{net} / A) \geq (f_y / f_u)(\gamma_{M2} / \gamma_{MO}) \qquad\qquad (5.3.)$$

Each section of the girder shall fulfil :

$$N_{tSd} \leq N_{tRd} \qquad\qquad (5.4.)$$

where N_{tSd} is the design value of the tensile force.

3.2. Resistance to compression force

The design compression resistance of a Class 4 cross-section is :

$$N_{cRd} = A_{eff} f_y / \gamma_{M1} \qquad\qquad (5.5.)$$

where A_{eff} is the cross-sectional effective area for compression, determined in accordance with previous Sections 2; γ_{M1} is taken equal to 1.10. Usually fastener holes need not be considered when computing A_{eff}.

When a Class 4 cross-section is not symmetrical and is subject to axial compression, there is a shift e_N of the centroïd of the effective section for compression compared to the one of the gross cross-section. The section shall thus be checked for combined compression and bending (see Section 3.4.).

Each section of the girder shall fulfil :

$$N_{cSd} \leq N_{cRd} \tag{5.6.}$$

where N_{cSd} is the design value of the compressive force.

3.3. Resistance to bending

Apart from shear force effects, the design moment resistance of a Class 4 cross-section *without bolt holes* is :

$$M_{Rd} = W_{eff} f_y / \gamma_{M1} \tag{5.7.}$$

where W_{eff} is the section modulus of the effective section for bending, computed in accordance with previous Lecture 2 and Section 2; γ_{M1} is taken equal to 1.10.

Usually fastener holes in the compression zone of the cross-section need not be allowed for. Fastener holes in the tension flange need not be allowed for provided that the condition (5.3.) on ductile behaviour be fulfilled; if not, a reduced tensile flange area, derived from this condition, must be assumed for the computation of W_{eff}. Fastener holes in the tension zone of the web need not be allowed for provided that the condition (5.3.) is satisfied for the complete tension zone composed of the tension flange plus the tension zone of the web.

Each section of the girder shall fulfil :

$$M_{Sd} \leq M_{Rd} \tag{5.8.}$$

where M_{Sd} is the design value of the bending moment.

3.4. Resistance to combined bending and axial (compressive) force

Apart from shear force effects, a Class 4 cross-section is satisfactory when the stresses computed based on the appropriate effective sections satisfy the appropriate criterion, i.e. the elastic criterion. This criterion writes :

$$(N_{Sd} / A_{eff}) + (M_{Sd} + N_{Sd} e_N) / W_{eff} \leq f_y / \gamma_{M1} \tag{5.9.}$$

where A_{eff} = the area of the effective cross-section for axial compression only;

W_{eff} = the section modulus of the effective section for bending only;

e_N = the shift of the centroïdal axis of the effective section for axial compression compared to the centroïdal axis of the gross section.

3.5. Shear buckling resistance

In accordance with the ULS design philosophy, the shear buckling resistance of a slender web is evaluated by accounting for the postcritical strength reserve. It depends on :

a) the depth-to-thickness ratio d / t_w ;
b) the spacing a of the possible transverse web stiffeners;
c) the efficiency of the anchor provided to the tension band by the flanges and the transverse stiffeners;
d) the presence of possible longitudinal stiffeners (not considered here).

The anchor provided by the flanges to the tension band is affected by the direct stresses present in these flanges, as a result of bending and/or axial load acting on the girder.

In the present state of knowledge, it is recommended to verify the shear buckling resistance using :

a) the *simple postcritical method* for unstiffened webs - under the reservation that the web has transverse stiffeners at the supports - or for webs fitted with largely spaced transverse stiffeners (spacing-to-depth ratio a/d larger than 3);
b) the *tension band method* for closely spaced transverse stiffeners $(1 \leq a / d \leq 3)$ provided adjacent web panels or stiff end posts provide anchor to the tension band.

The design value of the shear force V_{Sd} shall fulfil :

$$V_{Sd} \leq V_{Rd} \tag{5.10.}$$

where V_{Rd} is the design buckling resistance. When transversely stiffened webs, it is normally sufficient to refer to values of V_{Sd} at mid-length of the web panel under consideration; a web panel is limited by consecutive transverse stiffeners, because the latter shall be designed in order to remain straight up to the ultimate limit state of the girder.

The background of both methods has been developed in Lecture 4. It results however from a statistical evaluation of the test results that the appropriate specified partial safety factor $\gamma_{M1} = 1.10$ is reached provided that, the design buckling resistance be written as follows :

a) for the *simple postcritical method* :

$$V_{Rd} = V_{ba} = dt_w t_u / g_{M1} \tag{5.11.}$$

b) for the *tension band method* :

$$V_{Rd} = V_{bo} = (V_{cr} + 0.9 V_t) / g_{M1} \tag{5.12.}$$

with :

$$V_{cr} = dt_w \tau_{cr} \qquad (5.13.)$$

$$V_t = g t_w \sigma_t \sin\theta \qquad (5.14.)$$

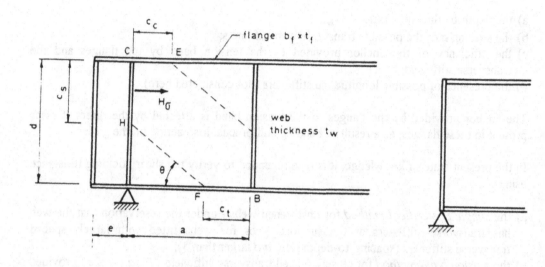

Figure 5.3 - Tension band in an end panel

End panels (fig. 5.3) of transversely stiffened girders must be considered with a special care. Indeed the web panel adjacent to an end post is only counterbalanced by one adjacent web panel only. Unless a suitable end post is supplied to anchor the tension field, the shear buckling resistance of the end panel shall be assessed using the simple postcritical method. An end post which resists the tension band force is termed *stiff end post* and shall satisfy the following criterion :

$$0.5 H_\sigma c_s \le M_{pC}^* + M_{pH}^* \qquad (5.15.)$$

H_σ is the horizontal component of the part of the resultant in the tension band, which anchors to the end post :

$$H_\sigma = \sigma_t (c_s \cos\theta - e \sin\theta) t_w \cos\theta \qquad (5.16.)$$

where e is the distance between the flanges of the H section constituting the end post; when a single plate end post is used, e is equal to zero.

The distance c_s between points C and H is computed as :

$$c_s = d - (a - c_t) tg\theta \qquad (5.17.)$$

Criterion (5.15.) is derived from the principle of virtual work applied to a plastic beam mechanism in the end post, with plastic hinges at C - top of the end post - and H; it means that the end post may not reach its ultimate plastic limit state before the end web panel has developed its shear buckling resistance. M^*_{pC} and M^*_{pH} are the plastic moments at points C and H, account taken of the influence of axial force existing at C and H respectively (see herebelow).

The shear buckling resistance of an end panel fitted with a *stiff* end post is computed as for an intermediate panel. However, because there is no possibility for self-equilibrium of the forces in the tension band, the bending capacities M^*_p used in the expression of c_e must account for axial forces N_f and N_s existing respectively in the flange and in the post, as a result of the anchor of the tension band. At point E, N_f is the horizontal component of the resultant force in the tension band :

$$N_{f,E} = g t_w \sigma_t \cos\theta \tag{5.18.}$$

At point C, the bending capacity is the lesser of two values : either the plastic bending resistance of the flange, reduced by the axial force :

$$N_{f,C} = c_s t_w \sigma_t \cos^2\theta \tag{5.19.}$$

or the plastic bending moment of the end post, reduced by the axial force :

$$N_{s,C} = c_c t_w \sigma_t \sin^2\theta \tag{5.20.}$$

At point H, the axial force in the end post is :

$$N_{s,H} = V_{Sd} - (d - c_s) t_w \tau_{cr} \tag{5.21.}$$

When a single plate end post is used, the reduced plastic moments write :

$$M^*_{pE} = 0.25 b_f t_f^2 f_{yf} [1 - (N_{f,E} / b_f t_f f_{yf})^2] \tag{5.22.}$$

$$M^*_{pC} = Min \begin{bmatrix} 0.25 b_f t_f^2 f_{yf} [1 - (N_{f,C} / b_f t_f f_{yf})^2] \\ 0.25 b_s t_s^2 f_{ys} [1 - (N_{s,C} / b_s t_s f_{ys})^2] \end{bmatrix} \tag{5.23.a,b}$$

$$M^*_{pH} = 0.25 b_s t_s^2 f_{ys} [1 - (N_{sH} / b_s t_s f_{ys})^2] \tag{5.24.}$$

with b_f : flange width;
       t_f : flange thickness;
       b_s : single plate end post width;
       t_s : single plate end post thickness;

f_{yf} : yield strength of the flange material;

f_{ys} : yield strength of the stiffener material.

A twin stiffener type of end post may be used as an alternative to the single plate type. It shall be noticed that the expressions of the reduced plastic moments of the end post shall be expressed accordingly.

The search for the shear buckling resistance of end panels is an iterative procedure; indeed the anchor length c_c depends on the unknown σ_t, which affects directly M_p^* values. As a first approximate, it is often assumed that the effect of axial force can be disregarded as far as it does not exceed 30 % of the squash load of the flange.

4. INTERACTION BETWEEN SHEAR, BENDING AND AXIAL FORCE

To which extent the tension band is able to anchor on the flanges depends on the flange stiffness; in the design rules, the latter is measured by the plastic bending resistance of the relevant flange. The theoretical shear buckling resistance shall thus depend on the relative direct stress magnitude in the flange. Many numerical simulations and tests have demonstrated that this interaction is rather limited; it is generally accepted that the design shear buckling resistance of the web needs not be reduced to allow for the moment and/or axial force, as long as the flanges can resist the whole of the design values of the bending moment and axial force.

When the latter condition is not fulfilled, the web has to contribute supporting direct stresses. That results in changes of the stress state over the web depth and from point to point in the web, on the one hand, and in a prematurous yielding of the web because of these additional direct stresses, on the other hand. It can be expected that the interaction between shear and direct stresses is more pronounced in this range. This fact is confirmed by test results.

Accordingly the interaction problem is summarized as follows (fig. 5.4.a,b) :

- The cross-section may be assumed to be satisfactory when both following criteria are fulfilled :

$$M_{Sd} \leq M_{fRd} \tag{5.25.}$$

and

$$V_{Sd} \leq V_{Rd}^o \tag{5.26.}$$

where M_{fRd} is the design plastic moment resistance of a cross-section consisting in the flanges only (taking possibly account of the effective width of the compression flange), with due allowance for a possible axial force in the relevant flange. The design shear resistance V_{Rd} is obtained from (5.11.) or (5.12.) depending on which one of the tension band model or the simple postcritical method is applicable; when use is made of (5.12.), V_t is computed conservatively with $c_c = c_t = 0$ and then $V_{Rd} = V_{Rd}^o$.

- Provided that V_{Sd} does not exceed 50 % of V_{Rd}^o, the design resistance of the cross-section to bending moment and axial force need not be reduced to allow for the shear force.
- When V_{Sd} exceeds 50 % of V_{Rd}^o, the following criterion should be satisfied :

$$M_{Sd} \leq M_{fRd} + (M_{NRd} - M_{fRd})[1 - (2V_{Sd}/V_{Rd}^o - 1)^2] \qquad (5.27.)$$

where M_{NRd} is the reduced plastic bending resistance allowing for the axial force N_{Sd}.

- The cross-section is able to resist any combination of internal forces (M_{Sd}, V_{Sd}) when the representative point lies within the boundaries of the interaction curve defined accordingly.

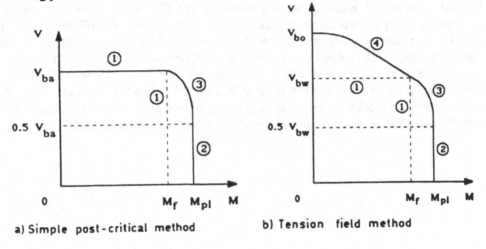

a) Simple post-critical method   b) Tension field method

Figure 5.4 - Interaction between shear and bending : Section of Class 4

5. INFLUENCE OF OUT-OF-PLANE DISPLACEMENTS ON SERVICE CONDITIONS

Redistribution of longitudinal direct stresses as well as the formation of a tension band in the web requires that appreciable out-of-plane displacements have occurred at the ultimate limit state. Because of the large web slenderness allowed for by exploiting the postcritical buckling strength reserve, the behaviour of slender webs in service conditions may be questionable.

The serviceability criteria linked to the magnitude of the out-of-plane displacements are mainly related to psychological considerations - i.e. to possible limits for visible buckles which might induce suspicion regarding the safety of the structure - and to potential snap-through phenomena - when initial out-of-flatness of the web is in sympathy with the plate buckling mode-. Such criteria are presently not given in codes. They are likely not to govern often the design.

On another hand out-of-plane displacements due to service loads may induce transverse bending stresses in the web plate, the magnitude of which would be likely to produce fatigue cracks at the web panel boundary, more especially in the vicinity of the compression flange. In contrast to what is generally assumed for the computations, the web is not actually hinged to the flanges so that the torsional stiffness of the latter offers a partial flexural restraint with resulting bending stresses in the fillet welds. Even small ranges of such stresses may induce fatigue cracks if the number of cycles is sufficiently high. This problem is presently being investigated by means of numerical simulations. A simple approach consisted in limiting the web slenderness; it was usually required that the depth of the compression zone of the web did not exceed 100 times the web thickness; for symmetrical plate girders subject to bending, this criterion gives : $d / t_w \leq 200$. Theoretical and experimental works are presently in progress in several countries; their results should provide with a better understanding of the phenomenon in a very near future.

LECTURE 6

COMPRESSION FLANGES OF BOX GIRDERS

1. INTRODUCTION

Box girder sections are employed to best advantage when use is made of their considerable *torsional stiffness*. Most of them have flanges which are wider and more slender than in plate girder construction, while the webs are usually of comparable slenderness.

The extensive use of slender plates makes stability considerations especially important in the box girder construction.

In present lecture, only aspects which are additional to those described previously are treated. Most of them are a consequence of the use of wide flanges.

First of all, the width of the flanges is likely to produce more or less pronounced *shear-lag*. The latter affects not only the distribution of the direct stresses within a specified cross-section but it interacts in addition with plate buckling of the compression flange.

Second, because of the slenderness of the flanges, the anchor of the tension band on the flanges becomes more questionable because the flanges cannot of course be fully mobilised for this purpose.

Third, the width of the compression flange may require that this flange be longitudinally and transversely stiffened. The behaviour of such a stiffened plated structure subject to predominant compression is a problem of major importance. In this respect, it is while stressing that resounding accidents occurred in the late sixties and early seventies have demonstrated that the design approaches used in the past were no more appropriate. These tragedies were the starting point of an intensive research work, which finally provided more realistic design approaches, because based on the ultimate strength of stiffened compression flanges, with due allowance made for initial geometric imperfections and residual stresses.

2. BUCKLING OF WIDE COMPRESSION FLANGES

Compression flanges of box girders may be made with a relatively thick plate, which is stable by its own. This solution which aims at the optimum cost - because it does not require stiffeners and avoid many welds - rather than at the minimum weight is quite appropriate for narrow box girders.

Most often the width of the compression flange is such that the latter needs to be stiffened in both longitudinal and transverse directions.

The unstiffened flange of a box girder can be treated as a plate by using a reduced effective width to account for the effects of plate buckling. When orthogonally stiffened, the flange is likely to exhibit different possible modes of buckling :

a) *overall buckling* of the orthogonally stiffened flange;
b) *buckling of the longitudinally stiffened panels* between transverse stiffeners;
c) *local buckling of the plate sheet* between longitudinal stiffeners;
d) *stiffener buckling*.

Of course two or more of these modes of buckling may combine and interact.

2.1. Limitations on the shape of longitudinal stiffeners

In order to prevent from torsional buckling or even local plate buckling of the longitudinal stiffeners, following requirements shall be fulfilled for the latter.

For *flat stiffeners* or *open section stiffeners*, the width-to-thickness ratio of any outstand shall comply with :

$$b/t_f \leq 12,5\,\varepsilon \qquad (6.1.)$$

with $\varepsilon = \sqrt{235/f_y}$, where f_y is expressed in N/mm^2.

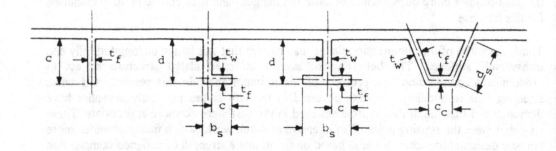

Figure 6.1 - Requirements for longitudinal stiffeners

The depth-to-thickness ratio d/t_w of the stiffener web shall also be limited. According to the *ECCS Recommendations (Publication n° 60)*, the relevant requirement is :

$$d/t_w \leq 30\varepsilon\left[1-1,25\sqrt{0,3-(1-\frac{a/b_s}{30\varepsilon})^2}\right] \qquad (6.2.)$$

where a is the spacing of the transverse stiffeners and b_s the width of the stiffener flange.

For *closed section stiffeners*, which are made with a cold formed steel sheet of thickness t, the width-to-thickness ratio of the flange and the web shall fulfill :

$$c_c / t \leq 40 \, \varepsilon \qquad\qquad\qquad\qquad\qquad\qquad\qquad (6.3.)$$
$$d_s / t \leq 40 \, \varepsilon \qquad\qquad\qquad\qquad\qquad\qquad\qquad (6.4.)$$

Above limitations are aimed at avoiding induced flange buckling, a phenomenon which has a non-ductile nature. When fulfilled, local or overall buckling of the longitudinal stiffeners properly will not precede the attainment of the ultimate strength of the flange.

2.2. Transverse stiffeners

Transverse stiffeners are usually designed with a view to provide nodal lines to the longitudinally stiffened panels. (Some of them are yet parts of diaphragms or bracings, aimed at retaining the box shape or distributing the forces to the support bracings).

To fulfil this role, transverse stiffeners of compression flanges shall satisfy two kinds of requirements :
a) *stiffness* requirements;
b) *strength* requirements.

When checking both, the section to be considered is the one composed of the transverse stiffener properly plus an associated effective width of plating. This effective width on each side of the stiffener web is, according to *E.C.C.S. Publication n° 60* :

   $a/4$ or $b/8$ for *strength* checks
   $a/2$ or $b/8$ for *stiffness* checks

where a and b are respectively the spacing of the transverse stiffeners and the width of the compression flange (measured between two successive webs).

Possible cut-outs of the stiffener web shall be allowed for, especially to permit the continuity of the longitudinal stiffeners throughout the transverse stiffener.

For the aforementioned checks, the transverse stiffener is supposed to be a simply supported beam fitted with an initial out-of-straighness of magnitude w_o equal to :

$$\frac{b}{300}, \frac{a}{300} \quad \text{or } 10 \, mm$$

according to which value is the lesser.

This beam shall carry the deviation forces from the adjacent compression panels under the assumption that both adjacent transverse stiffeners are rigid and are perfectly straight. The adjacent flange panels are assumed to be simply supported at the transverse stiffeners. The deviation forces are deduced from the longitudinal force in the flange and the above "differential bearing settlement".

It shall be verified that :

a) the maximum stress in the stiffener does not exceed the design material yield stress f_y / γ_{MO}, γ_{MO} being a partial safety factor.
b) the deflection is less than $b/300$.

2.3. Buckling strength

Provided above requirements be satisfied regarding the transverse stiffeners and the longitudinal stiffeners, the buckling of the stiffened compression flange is controlled in practice by the plate buckling of the longitudinal stiffened panel, accompanied or not by local plate buckling between the longitudinal stiffeners.

The elastic buckling stress of stiffened compression flanges used together with values of safety factors determined fiom test results was used in the past as a basis for design. It is now largely agreed that these elastic methods should be replaced by methods designed to predict the ultimate strength. In the frame of this lecture, only the latter is reviewed.

As far as inelastic buckling of stiffened compression flanges is concerned, three basic methods can be contemplated. In increasing degree of complexity, they are respectively :

a) the *column approach*;
b) the *orthotropic plate approach*;
c) the *discretely stiffened plate approach*.

2.3.1. Column approach

This procedure is simple in its concept, easy for its use and usually sufficiently accurate; it is therefore suitable for daily design purposes. It may be found significantly conservative where the sub-panel slenderness (plate sheet between longitudinal stiffeners) is low and the stiffener column slenderness high; in such situations, the orthotropic plate approach, which takes advantage of the postbuckling strength of the stiffened plate, fits much better with finite element predictions.

The *column approach* consists in treating the longitudinally stiffened flange as a series of unconnected struts, each of them being composed of a stiffener properly and a width of plate sheet equal to the spacing of the longitudinal stiffeners. Such a strut is then designed for column buckling.

Provided the transverse stiffeners of the compression flange be designed to remain stiff, the buckling length of these struts is taken equal to the spacing of the transverse stiffeners. That is however a safe approximate; indeed the strut has usually not the same critical buckling stress for buckling towards the plating or towards the stiffeners.

Allowance is made for reduction in effectiveness due to local buckling of the plate sheet between the stiffeners. That means that only an effective width of plate is associated to a stiffener (fig. 6.2).

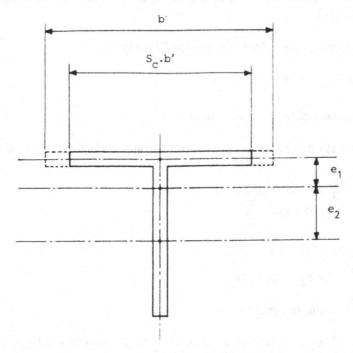

Figure 6.2 - Cross-section of a typical column

In addition, account is taken of the effect of initial distortions and residual stresses caused by the welding of the stiffeners onto the plate. Just one effective width may be used to account for both the reduced strength and stiffness of the compressed plate; doing so is just an approximation but has been found sufficient for design.

The E.C.C.S. approach is rather pragmatic. Each longitudinal stiffener is considered with its associated width of plating. Because the plating is likely to buckle between the transverse stiffeners, only an effective width of plating is considered for the assessment of the buckling resistance. Thus the cross-sectional area of one strut is :

$$A_{be} = A_s + S_c b' t \qquad (6.5.)$$

where :

A_s :   cross-sectional area of the stiffener properly;
b' :   spacing of the transverse stiffeners;
S_c :   effectiveness of the plate subpanel;

t :   plating thickness.

Factor S_c may be given by a Winterlike formula or an equivalent normalized plate buckling curve. Possibly it may be distinguished between restrained and unrestrained subpanels.

It is usual to consider that the compression flange involves not only the aforementioned struts but also two effective widths of plating $0.5\ S_c\ b'$, each of them being adjacent to one web of the box-girder.

Thus the cross-sectional area of the compression flange is :

$$A_f = m\ A_{be} + S_c\ b'\ t \tag{6.6.}$$

where m is the number of longitudinal stiffeners.

The ultimate strength σ_R of a strut is derived from an appropriate column buckling curve :

$$\overline{N}_R \equiv \frac{\sigma_R}{f_y^*} = \frac{1}{\phi + \sqrt{\phi^2 - \overline{\lambda}^2}} \leq 1 \tag{6.7.}$$

where : $\phi = 0,5\left[1 + \alpha^*(\overline{\lambda} - 0,2) + \overline{\lambda}^2\right]$ $\tag{6.8.}$

$\overline{\lambda} = \lambda / \lambda_E = \lambda / 93,9\varepsilon$

$\lambda = a / i$

i = radius of gyration

The strut is much likely to exhibit an initial out-of-straightness which is larger than the one implicitly accounted for ($1/1000$ of the buckling length) in the regular column buckling curves; an out-of-straightness of $a/500$, which would correspond to fabrication tolerance, shall be accounted for. It has been demonstrated by *Rondal and Maquoi* that such an account is taken by increasing the usual imperfection buckling factor α up to α^*, according as :

$$\alpha^* = \alpha + \frac{0.09}{i/e} \tag{6.9.}$$

where e is the distance between the reference fibre and the centroïd. As the strut is similar to a welded section, buckling curve c should be applied, i.e. $\alpha = 0.34$. Because its section is not symmetrical, one has to consider two values e_1 and e_2 of e, which correspond respectively to the mid-plane of the plating and to the centroïd of the stiffener properly. That means that two values of σ_R will be computed and the lesser $(\sigma_R)_{min}$ will be kept to check the resistance of the compression flange. The values of σ_R depend on the material yield stress. To account for possible coincident shear, a fictitious reduced value of the yield stress shall be used when computing σ_R with $e = e_1$; it is given by :

$$f_y^* = \sqrt{f_y^2 - 3\tau^2} \tag{6.10.}$$

where τ is an average shear stress in the flange, obtained from the superimposition of shear stress τ_t due to torsion and of (average) shear stress $(\tau_y/2)$ at the web-to-flange junction. When $e = e_2$, f_y^* is equal to f_y.

Last, account must be taken of the possible shear lag effect through factor ψ.

Finally, the buckling strength of the compression flange writes :

$$N_R = (m \, A_{be} + S_c \, b' \, t) \psi \, (\sigma_R)_{min} \tag{6.11.}$$

The question may arise to determine to which extent there is an interaction between shear lag and plate buckling. Indeed, should yielding occur, then the detrimental influence of elastic shear lag is reduced. Recent research has demonstrated that shear lag is not important at the ultimate limit state as long as certain geometric limits are imposed.

A first proposal has been the following :

a) $\psi = 1$ for $b / L \le 0.2$ (L being the girder span in the case of simply supported beams or the distance between points of contraflexure for continuous spans);
b) $\psi = \psi_e (b / 2L)$ for $b / L > 0.2$, where ψ_e is the elastic shear lag effective width ratio, which is given by formulae, charts or tables.

Later, based on a lot of numerical simulations, *Harding* suggested :

a) $\psi = 1$ for $(b / L) \le 0.2$;
b) $\psi = \psi_e (2,26 - \dfrac{\delta^2 \lambda^2}{13000}) > \psi_e$ for $0.2 \, b / L < 1$

where $\delta = m \, A_s / bt$ is the relative area of the stiffener and λ the slenderness ratio of the stiffener with account being taken of its effective cross-section.
c) for $(b / L) \ge 1$, ψ_e should be taken as the value for $(b / L) = 1$ divided by b/L.

2.3.2. Orthotropic plate approach

Compared to the column approach, the orthotropic plate approach is less conservative especially when light closely spaced stiffeners or for flanges where the longitudinally stiffened panel has a large aspect ratio. It is mathematically more complex.

In this approach, the discretely stiffened plate is replaced by two continuous structural systems : an isotropic flange and a notional plate which represents the longitudinal stiffeners. Because the stiffeners are one-sided, the centroïds of both systems do not coincide and due allowance shall be made for the relevant eccentricities when evaluating the deflections and the stresses in the stiffened plate.

The analytical procedure has been developed by *Jetteur et al*; its validity has been supported by a comparison with a lot of test results. The ultimate limit state requirements are based on limiting the maximum membrane stress along the unloaded edges and the bending stresses at the plate centre to yield. Of course an allowance for local buckling of the plating between the stiffeners is incorporated.

The buckling resistance of the compression flange is given as :

$$N_R = (m\,A_{be} + S_c\,b'\,t)\,\bar{\sigma}_{orth} \qquad (6.12.)$$

This expression looks like the one used in the column approach; the sole difference lies in the average ultimate stress $\bar{\sigma}_{orth}$ instead of $\bar{\sigma}_R$. The stress $\bar{\sigma}_{orth}$ is given as the minimum of $\bar{\sigma}_{oe}$, $\bar{\sigma}_{op}$ or $\bar{\sigma}_{os}$.

- $\bar{\sigma}_{oe}$ is the mean stress in the stiffened plate (not allowing for local plate buckling) corresponding to yield at the unloaded edge;
- $\bar{\sigma}_{op}$ is the mean stress corresponding to yield at the mid-plane of the plating at the centre of the stiffened plate including the effect of out-of-plane bending $(e = e'_1)$.
- $\bar{\sigma}_{os}$ is the mean stress corresponding to yield at the centroïd of the stiffener section $(e = e'_2)$.

The two values of fibre distance $(e'_1$ and $e'_2)$ should be used corresponding to the two directions of failure, i.e. buckling towards the stiffeners or towards the plating. Distance e'_1 is measured from the centroïd of the entire fully effective flange cross-section to the mid plane of the flange plate; distance e'_2 s measured from the same point to the centroïd of the stiffeners acting independently of the flange (fig. 6.3). It shall be noticed that e'_1 and e'_2 do not coincide with those, e_1 and e_2, used in Section 2.3.1.

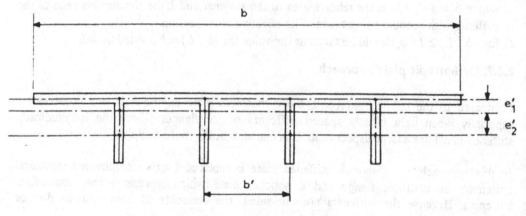

Figure 6.3 - Orthotropic plate

The stress σ_{oe} is given by equation as follows :

$$\bar{\sigma}_{oe} = 1.052\sqrt{\psi\sigma_{cr}f_y} - 0.277\sigma_{cr} \qquad (6.13.)$$

which is nothing else than the *Faulkner* formula where the effects of interaction between shear lag and plate buckling are introduced by means of factor ψ in the square root. Of course, the use of a Winterlike formula could also be contemplated. Shear lag factor ψ has been discussed by *Skaloud* in a separate set of lectures.

The critical buckling stress of the longitudinally stiffened plate is given by :

$$\sigma_{cr} = \frac{\pi^2\left[bD(1+\alpha^2)^2 + mEI + E(bt\,e_1'^2 + mA_s e_2'^2)\right]}{a^2(bt + mA_S)} \qquad (6.14.)$$

where : $D = Et^3 / 12(1-\nu^2)$:   plate bending stiffness;
          $\alpha = a / b$:   aspect ratio;
          I :   second moment of area of the stiffener alone;
          m :   number of stiffeners.

Stresses $\bar{\sigma}_{op}$ and $\bar{\sigma}_{os}$ are taken from figure 6.4 where e is either e_1' or e_2' according to the stress which is computed. The yield stress f_y^* to be used is defined in Section 2.3.1. for the two values of fibre distance e (here e_1' or e_2').

This stress is evaluated based on the assumption - mostly true in practice - that the stiffened plate fails in one half wave between the transverse stiffeners.

Whatever the approach - the column approach or the orthotropic plate approach - it shall be verified that the design load in the stiffened flange N_{sd} should not be less than $\bar{N}_R / \gamma_{M1}$, where γ_{M1} is an appropriate partial safety factor, taken equal to 1.1.

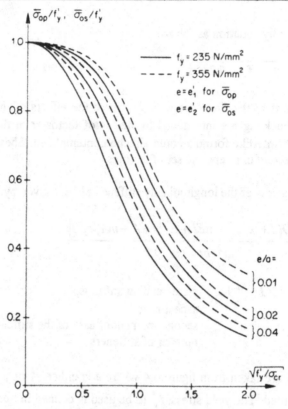

Figure 6.4 - Design charts

2.3.3. The discretely stiffened plate approach

First studies using this approach, based on elastic large deflection theory and defining the limit state of the plate as the onset of membrane yield, were conducted in Prague by *Skaloud et al* as early as the beginning of the sixties.

However, the real behaviour of a discretely stiffened plate involves the interactive inelastic buckling of plates and stiffeners, both locally and overall.

It is beyond the scope of normal design formulations to consider this interaction. At the present time only the most sophisticated non-linear computer programs (for example finite element or finite difference packages) can model this type of behaviour, and even these are at present too costly to carry out parametric studies.

As a result stiffener design rules are at present either based on elastic principles with stresses limited by the onset of yield or provide sufficient stiffener rigidity to preclude certain modes of buckling from the failure process. This latter approach is normally taken to avoid lateral torsional buckling or overall buckling of compression flanges including cross-frames.

Computer analyses have however been used to examine correlation with experimental data or provide general validation for design approaches.

Computer methods in general rely on the representation of the stiffeners as beam elements providing forces which can be incorporated with the plate analysis on an interactive basis. The yielding behaviour of the stiffener on a multiple-element basis and the plate on a single layer or multi-layer basis can and have been incorporated.

Computer analysis can now provide the solution to any specified problem but are not suitable for a general design method.

REFERENCES

1. E.C.C.S. : Behaviour and Design of Steel Plated Structures. Edited by P. Dubas and E. Gehri, Applied Statics and Steel Structures, Swiss Federal Institute of Technology, Zürich, Publication ECCS - Technical Working Group 8.3., n° 44, 1986.
2. Constructional Steel Design : an International Guide. Edited by P.J. Dowling, J.E. Harding and R. Bjorhovde. Elsevier Applied Science, 1992.
3. Djubek, J., Kodnar, R. and M. Skaloud : Limit State of the Plate Elements of Steel Structures. Birkhauser Verlag, Basel-Boston-Stuttgart, 1992.
4. C.E.N. : Eurocode 3 - Calcul des structures en acier - Partie 1.1. : Règles générales et règles pour les bâtiments. ENV 1993-1-1, July 1992.
5. E.C.C.S. : European Recommendations for the Design of Longitudinally Stiffened Webs and of Stiffened Compression Flanges. Publication n° 60, E.C.C.S., Brussels, 1990.

STEEL CONCRETE COMPOSITE CONSTRUCTION

T.M. Roberts
University of Wales Cardiff, Cardiff, UK

SUMMARY

The following six chapters are devoted to various aspects of the physical behaviour, analysis and detailed design of steel concrete composite construction.

Chapters 1 and 2 deal primarily with the ultimate and serviceability limit state design of composite steel beams and concrete slabs.

Following on from a brief introduction in chapter 2, chapter 3 presents a general theory of partial interaction. Applications of the theory considered include the nonlinear analysis of composite beams, the strengthening of reinforced concrete beams with adhesive bonded steel plates and the stress concentrations in adhesive joints.

The remaining chapters 4 to 6 deal with three less conventional forms of steel concrete composite construction, namely, composite plate girders with web cut outs, composite slabs with profiled steel sheeting and double skin composite construction.

CH 1   COMPOSITE ACTION AND ULTIMATE LIMIT STATE DESIGN

1.1   INTRODUCTION

Composite steel and concrete structures (Figure 1.1) are used extensively in the construction industry, to take advantage of the availability, cost and beneficial properties of the two materials, such as strength, workability and durability. One of the commonest forms of composite construction, illustrated in Figure 1.1(a), consists of a reinforced concrete slab, placed on top of and acting compositely with a series of steel beams or plate girders. The two materials are joined at their interfaces by so called "shear connectors", which transfer primarily shearing forces, and generally to a lesser extent normal forces, between the two materials. This form of construction utilises the tensile strength of the steel in conjunction with the compressive strength and void filling capability of the concrete. Also, during construction, the steel beams support the slab soffit formwork and wet concrete, thereby eliminating the need for temporary supports.

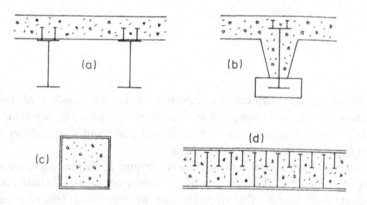

Figure 1.1   Composite sections (a) beam and slab (b) preflex beam (c) filled column
(d) double skin composite.

1.2   COMPOSITE ACTION

Composite action can be illustrated by considering the structures shown in Figures 1.2(a) and (b), each consisting of two rectangular beams of width b and depth d. The structure in Figure 1.2(a) has no shear connection at the interface and therefore acts as two independent beams. The structure shown in Figure 1.2(b) has "full" shear connection at the interface and therefore acts as a single beam of width b and depth 2d.

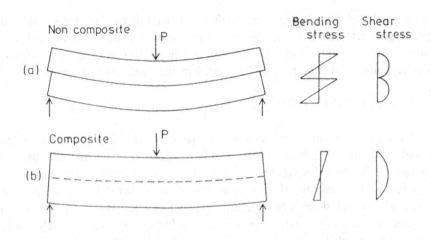

Figure 1.2   Composite action (a) non composite (b) fully composite.

The effective second moments of area of the two structures I_a and I_b, which are used to determine deflections (w ∝ 1 / E I) are :

$$I_a \quad = \quad 2 b d^3 / 12 \quad = \quad b d^3 / 6 \qquad\qquad (1.1)$$

$$I_b \quad = \quad b (2d)^3 / 12 \quad = \quad 4 b d^3 / 6 \qquad\qquad (1.2)$$

The effective section moduli Z_a and Z_b, which are used to determine bending stresses (f = M / Z) are :

$$Z_a \quad = \quad 2 b d^2 / 6 \quad = \quad b d^2 / 3 \qquad\qquad (1.3)$$

$$Z_b \quad = \quad b (2d)^2 / 6 \quad = \quad 2 b d^2 / 3 \qquad\qquad (1.4)$$

Finally, the section properties S_a and S_b, which are used to determine the maximum shear stresses (f = V S) are :

$$S_a = \frac{b\,(d/2)\,(d/4)}{2\,(b\,d^3/12)\,b} = \frac{3}{4\,b\,d} \qquad (1.5)$$

$$S_b = \frac{b\,d\,(d/2)}{(8\,b\,d^3/12)\,b} = \frac{3}{4\,b\,d} \qquad (1.6)$$

It can be concluded therefore that the maximum deflection and bending stress of the fully composite structure, are a quarter and a half respectively of those of the non composite structure. However, the maximum shear stresses in the two structures are equal, and in the fully composite structure have to be resisted by the shear connection.

1.3   SHEAR CONNECTION

Various forms of shear connector are used in composite construction, as illustrated in Figure 1.3, which can be broadly categorised as either rigid (bars and tees with hoops) or flexible (channels and headed studs). Apart from significant differences in shear stiffness and strength, rigid connectors cause higher stress concentrations and possible crushing and shear failure of the surrounding concrete. The failure modes of flexible connectors tend to be more consistent and ductile. The most widely used type of connector is the headed stud, which can be welded to the flanges of steel beams, either automatically or using portable hand tools.

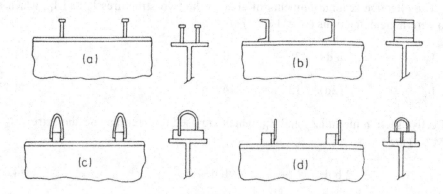

Figure 1.3   Shear connectors (a) headed stud (b) channel (c) bar and hoop
(d) T and hoop.

The characteristic shear resistance P_{Rk} and stiffness of stud connectors (static and fatigue) recommended in codes of practice, are normally based on extensive research and testing. Figure 1.4 illustrates a typical "push shear" test and the corresponding load-slip curve, from which the characteristic shear resistance, stiffness and ductility can be determined.

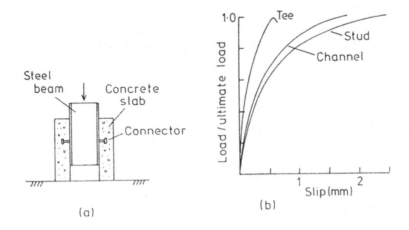

Figure 1.4   Testing of shear connectors (a) push shear test (b) load slip curves.

1.4   LIMIT STATE DESIGN

The basic principles of limit state design can be represented by the equations :

$$F_d \quad = \quad \gamma_F \, F_k \tag{1.7}$$

$$f_{Rd} \quad = \quad f_{Rk} \, / \, \gamma_m \tag{1.8}$$

$$f_{Rd} \quad \geq \quad f_d \tag{1.9}$$

Equation (1.7) relates the design actions or external forces F_d to the characteristic actions F_k multiplied by a load factor γ_F (typically 1.0 for serviceability limit states and 1.4 to 1.6 for ultimate limit states). Equation (1.8) expresses the design resistance (stress or stress resultant) of materials f_{Rd} in terms of the characteristic resistance f_{Rk} divided by a partial material safety factor γ_m (typically from 1.0 for steel to 1.5 for concrete). Finally, equation (1.9) specifies that the design resistance should be greater than or equal to the design internal force f_d corresponding to F_d.

In practice, composite sections are generally designed to satisfy the ultimate limit states, and serviceability limit states are checked subsequently by calculations based on assumed elastic behaviour. The remainder of this chapter deals with the ultimate limit state design of simply supported, composite beam and slab systems, illustrated in Figures 1.1(a) and 1.5.

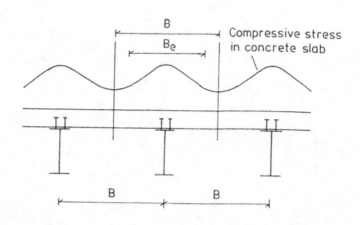

Figure 1.5   Effective width of a concrete slab.

1.5   ULTIMATE LIMIT STATE DESIGN OF COMPOSITE BEAM AND SLAB SYSTEMS

1.5.1   Effective width of concrete slab

In composite beam and slab systems (Figure 1.5) forces are transferred between the steel beams and concrete slab, along the line of the shear connection. Due to a phenomenon known as shear lag, this results in a non uniform distribution of compressive stress in the concrete slab, which is greatest along the line of the shear connection and least mid way between the beams. The ratio of greatest to least stress also varies along the span.

The system can still be idealised and designed as a series of isolated T-beams, by considering a reduced or effective width of concrete slab B_e to act compositely with each I-beam. Most codes of practice give very simple formulae for the calculation of effective widths. For example, Eurocode No 4 gives the maximum effective width of concrete slab, for simply supported systems in buildings, as one eighth of the span.

1.5.2   Ultimate moment resistance

Assumed plastic stress blocks in a composite beam and slab, at ultimate moment, are shown in Figure 1.6. Herein f_{ck} represents the compressive cylinder strength of the concrete. The three cases which need to be considered are when the plastic neutral axis is situated in (i) the concrete slab (ii) the top flange of the I-beam and (iii) the web of the I-beam.

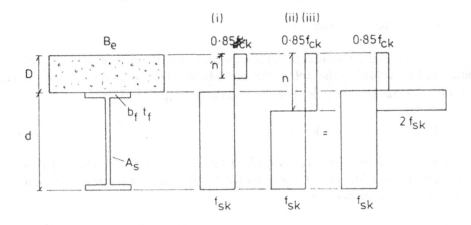

Figure 1.6   Assumed plastic stress blocks in a composite beam and slab.

(i)   Plastic neutral axis in concrete slab

Equating the resultant compressive and tensile forces acting within the section gives :

$$B_e \, n \, 0.85 \, f_{ck} / \gamma_c \quad = \quad A_s \, f_{sk} / \gamma_s \tag{1.10}$$

Taking moments about the centroid of the concrete stress block gives the design resistance moment M_{Rd} as :

$$M_{Rd} \quad = \quad A_s \, (D + d / 2 - n / 2) \, f_{sk} / \gamma_s \tag{1.11}$$

(ii)   Plastic neutral axis in top flange of I-beam

$$B_e \, D \, 0.85 \, f_{ck} / \gamma_c \quad = \quad \{A_s - 2 \, b_f (n - D)\} \, f_{sk} / \gamma_s \tag{1.12}$$

$$M_{Rd} \quad = \quad \{A_s\,(D + d)\,/\,2 - 2\,b_f\,(n - D)\,n\,/\,2\}\,f_{sk}\,/\,\gamma_s \tag{1.13}$$

(iii)   Plastic neutral axis in web of I-beam

$$B_e\,D\,0.85\,f_{ck}\,/\,\gamma_c \quad = \quad \{A_s - 2\,b_f\,t_f - 2\,t_w\,(n - D - t_f)\}\,f_{sk}\,/\,\gamma_s \tag{1.14}$$

$$M_{Rd} \quad = \quad \{A_s\,(D + d)\,/\,2 - 2\,b_f\,t_f\,(D + t_f)\,/\,2$$

$$- 2\,t_w\,(n - D - t_f)\,(n + t_f)\,/\,2\}\,f_{sk}\,/\,\gamma_s \tag{1.15}$$

1.5.3   Ultimate shear resistance

The design shear resistance V_{Rd} of a composite beam and slab is generally assumed to be provided by the web of the steel I-beam. For stocky webs which are not susceptible to instability, the yield shear stress is equal to $f_{sk}\,/\,\sqrt{3}$. Therefore :

$$V_{Rd} \quad = \quad d_w\,t_w\,f_{sk}\,/\,\sqrt{3}\,\gamma_s \tag{1.16}$$

For slender webs which are susceptible to instability, a reduced value of f_{sk} can be incorporated in equation (1.16). Alternatively, the design shear resistance can be determined from tension field theory.

1.5.4   Ultimate connector resistance

The number of connectors n_c between points of zero and maximum moment, should be capable of resisting the ultimate compressive and tensile forces, acting within the cross-section at the point of maximum moment. Hence :

$$n_c\,P_{Rk}\,/\,\gamma_v \quad \geq \quad B_e\,D\,f_{ck}\,/\,\gamma_c$$

$$\geq \quad A_s\,f_{sk}\,/\,\gamma_s \tag{1.17}$$

1.6   NOTATION

A   area
b , B   width
d   depth   ;   depth of steel beam
D   depth of concrete
E   Young's modulus
f   stress or stress resultant
F   load or action
I   second moment of area
M   moment
n   depth of neutral axis   ;   number of connectors
P   connector shear force
S   section property for calculating shear stress
t   thickness
V   shear force
w   displacement
Z   section modulus
γ partial safety factor for loads and materials

Subscripts

a , b   different components or materials
c   concrete   ;   connector
d   design
e   effective
f   flange
F   load or action
k   characteristic
m   material
R   resistance
s   steel
v   connector
w   web

1.7   BIBLIOGRAPHY

1.   Yam L C P. Design of Composite Steel Concrete Structures. Surrey University Press, London, 1981.

2.   Johnson R P. Composite Structures of Steel and Concrete : Vol 1. Second Edition, Blackwell Scientific Publications, 1994.

3.   Johnson R P and Buckby R J. Composite Structures of Steel and Concrete : Vol 2 : Bridges. Second Edition, Collins, London, 1986.

4.   Knowles P R. Composite Steel and Concrete Construction. Butterworths, London, 1973.

5.   Eurocode No 4 : Design of composite steel and concrete structures : Part 1.1 : General rules and rules for buildings. 1992.

6.   BS5400: Steel concrete and composite bridges : Part 5 : Design of composite bridges. British Standards Institution, 1979.

7.   Johnson R P and Anderson D. Designers' Handbook to Eurocode 4 : Part 1.1 : Design of Composite Steel and Concrete Structures. Thomas Telford, London, 1993.

CH 2 COMPOSITE ACTION AND SERVICEABILITY LIMIT STATES

2.1 INTRODUCTION

One of the most significant advantages of the use of composite beam and slab systems is the ease and speed of construction, during which the steel beams support the slab soffit formwork and weight of wet concrete. This avoids the need for costly temporary supports and enables work to continue beneath. However, the sequence of construction may introduce stress discontinuities in the composite system, which influence its performance under service conditions. Various factors which influence the performance of composite beam and slab systems under service conditions are considered herein, including the construction sequence, partial shear connection, shrinkage creep and thermal effects, and fatigue of the shear connection.

2.2 ELASTIC ANALYSIS OF COMPOSITE SECTIONS

2.2.1 Full interaction

For a steel beam and concrete slab to act compositely, the connectors must possess adequate stiffness and strength. In reality, all connectors have finite stiffness and therefore slip must occur between the two materials, if the interactive shearing forces are to be developed. In many practical situations, slip and its influence on the stresses acting within the composite system are of relatively minor significance, and can be neglected in analysis and design.

For full interaction (negligible slip) the distribution of strain over the depth of a composite beam and slab, assuming plane sections remain plane, is as shown in Figure 2.1. Denoting the curvature of the composite system by K (inverse of the radius of curvature) and assuming that the concrete remains uncracked, the compressive and tensile forces in the concrete slab and steel beam are given by :

$$C \quad = \quad K E_c A_c (n - D / 2) \tag{2.1}$$

$$T \quad = \quad K E_s A_s (D + d / 2 - n) \tag{2.2}$$

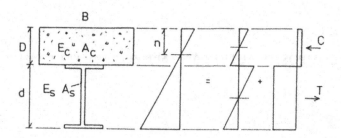

Figure 2.1   Strain distribution in a fully composite beam.

Equating compressive and tensile forces gives the depth of the neutral axis n as :

$$n = \frac{D}{2} + \frac{(D + d) E_s A_s}{2 (E_s A_s + E_c A_c)} \qquad (2.3)$$

The moment M acting within the composite section is given by :

$$M = K (E_c I_c + E_s I_s) + C (D + d) / 2 \qquad (2.4)$$

Substituting equations (2.1) and (2.3), M can be expressed as :

$$M = K (E_c I_c + E_s I_s) (1 + \alpha) \qquad (2.5)$$

$$\alpha = \frac{E_c A_c E_s A_s (D + d)^2}{4 (E_c A_c + E_s A_s) (E_c I_c + E_s I_s)} \qquad (2.6)$$

The bending stress f at distance z from the composite centroidal axis (f = K E z) is therefore given by :

$$f = \frac{M E z}{(E_c I_c + E_s I_s) (1 + \alpha)} \qquad (2.7)$$

in which E takes the value of E_c or E_s, depending upon whether the point under consideration lies within the concrete or steel.

The shear force per unit length q, at the interface between the concrete and steel (Figure 2.2(b)) is given by :

$$q \quad = \quad - dC \,/\, dx \qquad\qquad (2.8)$$

Hence, substituting C from equation (2.4) and K from equation (2.5), and noting that the transverse shear force V = - dM / dx (Figure 2.2(a)) gives :

$$q \quad = \quad \frac{2\,\alpha\,V}{(1+\alpha)\,(D+d)} \qquad\qquad (2.9)$$

The bending stresses and interface shearing forces in a fully composite section, can therefore be determined simply from equations (2.7) and (2.9).

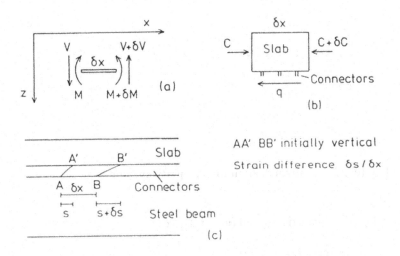

Figure 2.2   Interface slip and shearing forces (a) sign convention for shear force and bending moment (b) interface shear force (c) interface slip.

2.2.2   Partial interaction

Slip s at the interface between the concrete and steel induces a strain difference e between the two materials (Figure 2.2(c)) given by :

$$e \quad = \quad ds / dx \tag{2.10}$$

The distribution of strain over the depth of a partially composite section is therefore as shown in Figure 2.3, from which the internal forces and strain difference can be determined as :

$$C \quad = \quad K E_c A_c (n_c - D / 2) \tag{2.11}$$

$$T \quad = \quad K E_s A_s (d / 2 - n_s) \tag{2.12}$$

$$M \quad = \quad K (E_c I_c + E_s I_s) + C (D + d) / 2 \tag{2.13}$$

$$e \quad = \quad K (D + n_s - n_c) \tag{2.14}$$

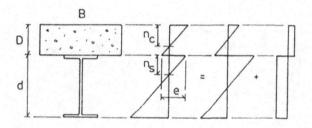

Figure 2.3   Strain distribution in a partially composite section.

Equating C and T and eliminating K and $(n_s - n_c)$ gives :

$$M \quad = \quad \frac{(D + d) (1 + \alpha) C}{2 \alpha} \quad + \quad \frac{2 (E_c I_c + E_s I_s) e}{(D + d)} \tag{2.15}$$

Denoting the shear stiffness of the connectors per unit length by k_s, the interface shear force per unit length q can be expressed as :

$$q \quad = \quad k_s s \tag{2.16}$$

Hence from equations (2.10), (2.16) and (2.8) :

$$e \quad = \quad -(1/k_s)\, d^2C / dx^2 \tag{2.17}$$

Substituting equation (2.17) into equation (2.15) gives the governing differential equation for C as :

$$\frac{2(E_c I_c + E_s I_s)}{k_s (D+d)} \cdot \frac{d^2C}{dx^2} - \frac{(D+d)(1+\alpha)}{2\alpha} \cdot C \quad = \quad M \tag{2.18}$$

Equation (2.18) can be solved analytically for simple loading and boundary conditions. Some solutions for simply supported beams are shown graphically in Figure 2.4. In general, full interaction theory provides an upper bound for the interface shearing forces q, which justifies the use of full interaction theory in practical design.

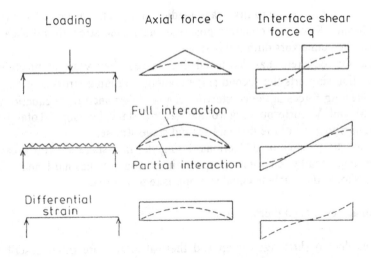

Figure 2.4   Variation of concrete compressive force C and interface shear force q in simply supported beams.

2.3   SERVICEABILITY LIMIT STATE DESIGN

2.3.1   Construction sequence

The construction and loading sequence for a composite steel beam and concrete slab, can be broadly divided into two main stages :

a)   Erect beam and place wet concrete
b)   Apply surfacing and imposed loading

During the first stage the steel beam supports the weight of the beam, slab soffit formwork, wet concrete and the construction loading (concreting plant and personnel). Bending stresses induced during the first stage are confined to the steel beam and can be calculated from the equation :

$$f_s = M z / I_s \qquad (2.19)$$

in which M is the moment due to first stage loading and y is the distance from the centroid of the steel beam. Since wet concrete possesses negligible strength, no shearing forces are induced in the connectors during this stage.

When the concrete has hardened the system behaves compositely and the composite section supports the second stage loading. Bending stresses in the composite section and shearing forces in the connectors can be determined from equations (2.7) and (2.9), with M and V corresponding to the second stage loading. Total stresses are obtained by superposition of the first and second stage stresses.

If either the stresses or deflections during the first stage are too great, the steel beam can be supported by temporary props until the concrete has hardened. The second stage loading should then include equal and opposite prop forces.

2.3.2   Shrinkage creep and thermal effects

Strains due to shrinkage, creep and thermal effects are often described as free strains, since when unrestrained they do not induce stress. In general the analysis of the influence of free strains in a composite section is complex and necessitates the application of partial interaction theory. However, approximate solutions for direct stresses can be obtained assuming full interaction.

The relatively simple case of a composite beam and slab, with a uniform free strain e_f (tensile positive) in the concrete slab (Figure 2.5) is used as an illustration. The free strains can be completely restrained by applying a force C_f to the concrete given by :

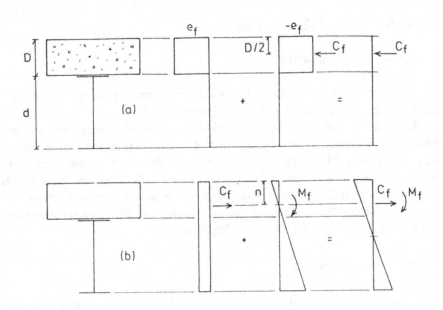

Figure 2.5   Free strain in a fully composite beam (a) restraint of free strain
(b) removal of restraining force.

$$C_f \quad = \quad E_c \, A_c \, e_f \qquad\qquad (2.20)$$

The corresponding moment about the centroid of the composite section is :

$$M_f \quad = \quad E_c \, A_c \, e_f \, (n - D / 2) \qquad\qquad (2.21)$$

where n is the depth of the fully composite neutral axis.

Since C_f does not exist in reality, it can be removed by applying an equal and opposite force to the composite section. Stresses throughout the composite section can then be determined by superposition of direct and bending stresses, induced by restraint of the free strains and subsequent removal of the restraining force.

Considering firstly the concrete stress and noting that the area of the transformed concrete section is $A_c + A_s \, E_s / E_c$, the total concrete stress (compression positive) is given by :

$$f_c \; = \; \frac{C_f}{A_c} \; - \; \frac{C_f}{(A_c + A_s \, E_s / E_c)} \; + \; \frac{M_f \, E_c \, z}{(E_c \, I_c + E_s \, I_s)(1 + \alpha)} \qquad\qquad (2.22)$$

$$\log N \quad = \quad 22.32 \quad - \quad 8 \log f_d \tag{2.26}$$

2.3.4   Serviceability limit states

The performance of composite structures under service conditions is related primarily to deflections, stresses and fatigue (section 2.3.3).

Limiting deflections is intended primarily to avoid damage to non structural elements such as surfacing and cladding. It also ensures that structures possess adequate stiffness to avoid public concern regarding noticeable deformations and undesirable dynamic responses. Calculation of deflections is generally based on linear (elastic) full interaction analysis, taking account of the construction sequence as appropriate.

Allowable service stresses are generally limited by the characteristic strengths of materials divided by the partial material safety factors :

$$f_d \quad \leq \quad f_{Rk} / \gamma_m \tag{2.27}$$

Use of equation (2.27) requires that the structure possesses an adequate reserve of strength due to the spread of plastic strains. For example, the fully plastic moment of a rectangular steel beam is 50% higher than the moment corresponding to first yield.

Individual stud connectors do not however possess a post yield reserve of strength and may be adversely affected by repeated loading. Therefore, BS5400 gives the design shear load per connector under service conditions as :

$$P_d \quad \leq \quad 0.675\, P_{Rk} / \gamma_v \tag{2.28}$$

in which γ_v is assumed equal to 1.25.

The actual compressive force in the concrete C (area of concrete multiplied by stress at centroid) is therefore :

$$C = C_f - \frac{C_f E_c A_c}{E_c A_c + E_s A_s} - \frac{M_f E_c (n - D/2) A_c}{(E_c I_c + E_s I_s)(1 + \alpha)} \qquad (2.23)$$

Similarly, stresses in the steel beam (tensile positive) are given by :

$$f_s = \frac{C_f}{A_s + A_c E_c / E_s} - \frac{M_f E_s z}{(E_c I_c + E_s I_s)(1 + \alpha)} \qquad (2.24)$$

The preceding analysis provides no information concerning the forces in the shear connection, since the uniform moment M_f does not induce any transverse shear force (see equation (2.9)). Therefore, most codes of practice recommend approximate solutions for the interface shear, based on a uniform distribution of the concrete force C (equation (2.23)) over a specified length of the beam near the supports.

However, the forces in the shear connection and stresses throughout the composite section can be determined from partial interaction theory. Solution of equation (2.18) subject to the boundary conditions $C = - C_f$ at the free ends, enables the distribution of C and q along the beam to be determined. Such solutions indicate that q increases rapidly towards the ends of simply supported beams, as shown in Figure 2.4.

2.3.3   Fatigue of shear connection

Composite beam and slab construction is used widely for short and medium span bridges. Elements of such bridges, in particular the shear connection, are subjected to repeated loading which may induce premature fatigue failure.

Fatigue resistant design is generally based on S-N curves which relate an actual or nominal stress range S, to the number of cycles to fatigue failure N. BS5400 defines the nominal design stress range for a welded stud connector f_d as :

$$f_d = \frac{425 P_d}{P_{Rk} / \gamma_v} \qquad (N / mm2) \qquad (2.25)$$

in which P_d is the design shear load range applied to the stud under service conditions, P_{Rk} is the characteristic stud resistance and γ_v is the partial material safety factor. The number of cycles to fatigue failure at a stress range f_d is then given by the equation :

2.4   NOTATION

A	area
C	compressive force
d	depth of steel beam
D	depth of concrete
e	strain ; strain difference
E	Young's modulus
f	stress
I	second moment of area
k	stiffness per unit length
K	curvature
M	moment
n	depth of neutral axis
N	number of cycles
P	connector shear force
q	shear force per unit length
s	slip
S	stress range
T	tensile force
V	shear force
z	distance from centroid
γ	partial safety factor for loads and materials

Subscripts

c	concrete
d	design
f	free
k	characteristic
m	material
R	resistance
v	connector

2.5   BIBLIOGRAPHY

1.   Yam L C P. Design of Composite Steel Concrete Structures. Surrey University Press, London, 1981.

2.   Johnson R P. Composite Structures of Steel and Concrete : Vol 1. Second Edition, Blackwell Scientific Publications, 1994.

3.   Johnson R P and Buckby R J. Composite Structures of Steel and Concrete : Vol 2 : Bridges. Second Edition, Collins, London, 1986.

4.   Knowles P R. Comcposite Steel and Concrete Construction. Butterworths, London, 1973.

5.   Eurocode No 4 : Design of composite steel and concrete structures : Part 1.1 : General rules and rules for buildings. 1992.

6.   BS5400 : Steel concrete and composite bridges : Part 5 : Design of composite bridges. British Standards Institution, 1979.

7.   Johnson R P and Anderson D. Designers' Handbook to Eurocode 4 : Part 1.1 : Design of Composite Steel and Concrete Structures. Thomas Telford, London, 1993.

CH 3   PARTIAL INTERACTION THEORY AND APPLICATIONS

3.1   INTRODUCTION

The structural performance of composite systems depends upon the ability of the connections to transfer shear and normal forces between the various components.   In general, all connections possess finite stiffness.   Therefore, finite slip and separation must occur for the shear and normal forces to be fully mobilised.   This in turn may influence not only the forces in the main components of the composite system, but also the forces and stresses in the connection.

For many common forms of composite construction, such as steel beam and slab flooring systems and bridges, the influence of slip and separation is of relatively minor significance and can be neglected for most practical purposes.   However, full interaction theory does not provide any realistic information concerning either the influence of free strains (thermal, shrinkage etc.) or the stress concentrations in adhesive joints, which may have a significant influence on the performance of the composite system.   Realistic analysis of such problems requires the development and application of a general theory of partial interaction.

3.2   THEORY

An element of a two component composite system, length δx, is shown in Figure 3.1.   The system consists of two materials a and b, joined by a medium of assumed negligible thickness but having finite shear and normal stiffnesses.   Assuming plane sections within each material remain plane, the strains ε at depth z can be expressed in terms of displacements u and w as :

$$\varepsilon_a = u_{a,x} - z_a w_{a,xx} \tag{3.1}$$

$$\varepsilon_b = u_{b,x} - z_b w_{b,xx} \tag{3.2}$$

in which subscripts x denote differentiation.

Stresses σ are related to strains and material properties E (linear or nonlinear) by the equations :

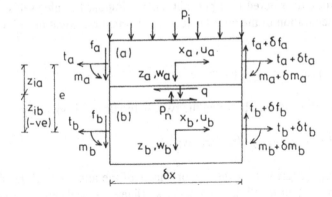

Figure 3.1   Element of a composite beam.

$$\sigma_a = E_a (\varepsilon_a - \varepsilon_{fa}) \tag{3.3}$$

$$\sigma_b = E_b (\varepsilon_b - \varepsilon_{fb}) \tag{3.4}$$

in which ε_f denotes free strain due to temperature, creep etc.

The axial forces t and moments m are obtained by integrating the stresses over the cross section areas. Hence :

$$t_a = \int E_a (u_{a,x} - z_a w_{a,xx} - \varepsilon_{fa}) \, dA_a \tag{3.5}$$

$$t_b = \int E_b (u_{b,x} - z_b w_{b,xx} - \varepsilon_{fb}) \, dA_b \tag{3.6}$$

$$m_a = - \int E_a (u_{a,x} - z_a w_{a,xx} - \varepsilon_{fa}) z_a \, dA_a \tag{3.7}$$

$$m_b = - \int E_b (u_{b,x} - z_b w_{b,xx} - \varepsilon_{fb}) z_b \, dA_b \tag{3.8}$$

For equilibrium of the element in the x - direction :

$$t_{a,x} + t_{b,x} = 0 \tag{3.9}$$

For equilibrium in the z - direction and moment equilibrium :

$$f_{a,x} + f_{b,x} = p_t = p_i + p_a + p_b \tag{3.10}$$

$$m_{a,x} + m_{b,x} = f_a + f_b + e \, t_{b,x} \tag{3.11}$$

where p_t is the total distributed load per unit length (imposed p_i plus self weights p_a and p_b) and e is the separation of the coordinate axes. Combining equations (3.10) and (3.11) gives :

$$m_{a,xx} + m_{b,xx} - e\, t_{b,xx} - p_t = 0 \qquad (3.12)$$

The slip at the interface between the two materials u_{ab} is given by :

$$u_{ab} = (u_a - z_{ia}\, w_{a,x}) - (u_b - z_{ib}\, w_{b,x}) \qquad (3.13)$$

in which z_{ia} and z_{ib} (negative) are the z coordinates of the interface. If the shear stiffness of the connection per unit length is denoted by k_s (linear or nonlinear) the interface shear force per unit length q is given by :

$$q = k_s\, u_{ab} = t_{a,x} \qquad (3.14)$$

Hence, substituting from equation (3.13) gives :

$$t_{a,x} - k_s \{(u_a - z_{ia}\, w_{a,x}) - (u_b - z_{ib}\, w_{b,x})\} = 0 \qquad (3.15)$$

If the normal stiffness of the connection per unit length is denoted by k_n (linear or nonlinear) the interface normal force p_n is given by :

$$p_n = k_n\, (w_b - w_a) \qquad (3.16)$$

For equilibrium of the element of material a in the z direction and for moment equilibrium :

$$f_{a,x} = p_i + p_a + p_n \qquad (3.17)$$

$$m_{a,x} + t_{a,x}\, z_{ia} = f_a \qquad (3.18)$$

Hence from equations (3.16) to (3.18) :

$$m_{a,xx} - t_{a,xx}\, z_{ia} - k_n\, (w_b - w_a) - p_i - p_a = 0 \qquad (3.19)$$

Equations (3.9), (3.12), (3.15) and (3.19) are the governing differential equations of equilibrium, which can be expressed in terms of the unknown displacements u_a, w_a, u_b and w_b by substituting equations (3.5) to (3.8). Simultaneous solution of these equations subject to prescribed boundary conditions determines the unknown displacements and hence the forces in materials a and b. The interface shear and normal forces can also be determined from equations (3.5) to (3.8) and (3.14) and (3.16).

In general the solution of the four simultaneous differential equations of equilibrium is lengthy and complex, and analytical solutions can only be obtained for simple loading and boundary conditions. For more complex situations and nonlinear material behaviour, the equations can be solved numerically using finite difference techniques. Examples of the application of these equations are given in following sections.

3.3   APPLICATIONS

3.3.1   Nonlinear analysis of composite beams

Figure 3.2(a) shows the cross-section of a composite steel beam and concrete slab, assumed to be simply supported over a span of 9 m and subjected to a uniformly distributed imposed load. The assumed stress strain characteristics of the concrete and steel and load slip characteristic of the shear connection, are shown in Figures 3.2(b) to (d). The normal stiffness of the connection was assumed constant and large enough to prevent any significant separation.

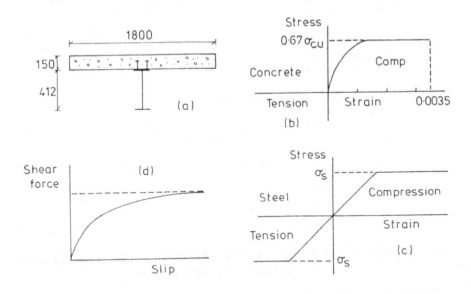

Figure 3.2   Composite beam and slab (a) cross - section dimensions (b) stress - strain relationship for concrete (c) stress - strain relationship for steel (d) force - slip relationship for shear connection.

The nonlinear load displacement response was determined using a nonlinear (iterative) finite difference solution of the governing differential equations derived in section 3.2. Since the assumed material properties were nonlinear, integration of equations (3.5) to (3.8) was performed numerically by dividing the concrete slab into 10 layers and the steel beam into 18 layers (10 for the web and 4 for each flange).

Results of the analysis are presented graphically in Figure 3.3. Figure 3.3(a) shows the nonlinear load displacement response, which converges approximately to the ultimate load p_u based on assumed fully plastic stress blocks.

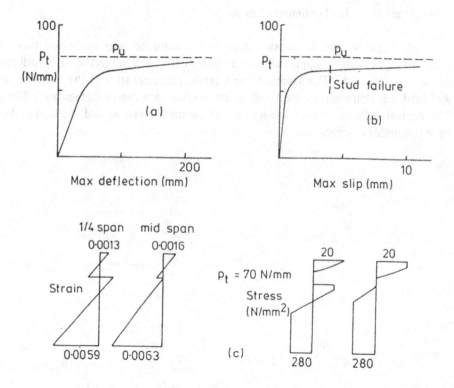

Figure 3.3   Composite beam and slab (a) load - deflection curve (b) load - slip curve (c) strain and stress profiles.

Figure 3.3(b) shows the load versus maximum (end) slip, which also converges approximately to p_u for relatively large slip. From this plot it is possible to determine the load at which the maximum slip capacity of the shear connection, based on push shear tests, is exceeded.

Strain and stress profiles throughout the depth of the composite beam, corresponding to $p_u = 60$ and 70 kN/m, are shown in Figure 3.3(c). The discontinuous strain profiles indicate the existence of slip at the interface between the concrete and steel, while the stress profiles indicate the spread of plasticity.

3.3.2 Strengthening of reinforced concrete beams with adhesive bonded steel plates

Since the early work in South Africa and France, the technique of strengthening concrete structures by bonding steel plates to the concrete surfaces with epoxy resins, has been used in various parts of the world. The advantages of the technique are that the work can be carried out relatively simply and quickly, even while the structure is still in use, with negligible changes in member dimensions.

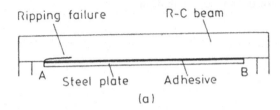

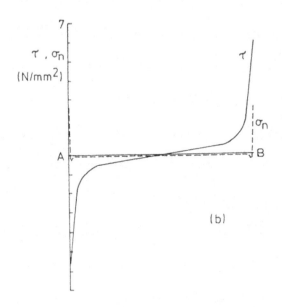

Figure 3.4   Concrete beam with adhesive bonded steel plate (a) ripping failure (b) stress concentrations in the adhesive layer.

   Systematic experimental investigations of the structural performance of plated concrete beams have indicated the possibility of "ripping" failures associated with high shear and normal stresses at the free ends of the plates, as illustrated in Figure 3.4(a). Theoretical predictions of the shear and normal stresses in the adhesive layer can be obtained from analytical solutions of the differential equations of equilibrium derived in section 3.2, assuming linear elastic material behaviour.

   Figure 3.4(b) shows typical results for the shear and normal stresses in the adhesive layer of a simply supported, reinforced concrete beam, subjected to uniformly distributed loading. As can be seen, both the shear and normal stresses increase rapidly towards the free ends of the plate. The physical explanation for the stress concentration is the stress discontinuity which exists at the free end of the adhesive layer, where the maximum shear stress (corresponding to maximum slip) reduces rapidly to zero. In reality, the high stress concentrations often result in nonlinear material behaviour and associated modification of the stress distributions.

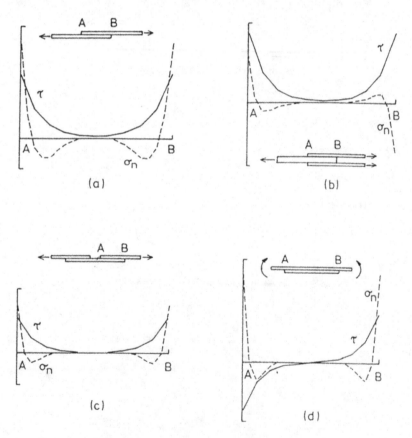

Figure 3.5   Stress concentrations in adhesive joints (a) single lap joint (b) double lap joint (c) strap joint (d) reinforced plate.

3.3.3   Stress concentrations in adhesive joints

The use of adhesives to manufacture force transmitting joints between structural materials is well established. In the aircraft industry adhesives are often used in place of welds and rivets, to reduce residual stresses and stress concentrations which may induce fatigue failures. Adhesives have also been used to attach cover plates to the flanges of steel beams, to avoid fatigue failures associated with welding. However, the structural performance of adhesive joints is influenced by the shear and normal stress concentrations in the adhesive layer. Theoretical predictions of these stresses can be obtained from analytical solutions of the differential equations derived in section 3.2.

Figures 3.5(a) to (d) show typical results for the shear and normal stresses in the adhesive layers of various forms of adhesive joint. In all cases the shear and normal stresses increase rapidly towards the free ends of the adhesive layer, which has a significant influence on the structural performance of the joint.

3.4   NOTATION

A	area
e	separation of local coordinate axes
E	Young's modulus
f	shear force
k	stiffness per unit length
m	moment
p	distributed force per unit length
q	shear force per unit length
t	tensile force
u , w	displacements in x and z - directions
x , z	dimensions and coordinate axes
ε	direct strain
σ , τ	direct and shear stresses

Subscripts

a , b	materials a and b
c	concrete
f	free
i	interface
n	normal
s	steel ; shear
t	total
u	ultimate
x	differentiation with respect to x.

3.5 BIBLIOGRAPHY

1. Goland M and Reissner E. The stresses in cemented lap joints. Journal of Applied Mechanics, Vol 11, 1944, pp A17-A27.

2. Newmark N M, Seiss C P and Viest I M. Tests and analysis of composite beams with incomplete interaction. Proc Soc Expt Stress Analysis, 9(1), 1951, pp 75-92.

3. Adams R D and Wake W C. Structural Adhesive Joints in Engineering. Elsevier Applied Science Publishers Ltd., 1984.

4. Roberts T M. Finite difference analysis of composite beams with partial interaction. Computers and Structures, Vol 21, No 3, 1985, pp 469-473.

5. Roberts T M and Haji - Kazemi H. Theoretical study of the behaviour of reinforced concrete beams strenghtened by externally bonded steel plates. Proc Instn Civ Engrs, Part 2, 87, March 1989, pp 39-55.

6. Roberts T M. Approximate analysis of shear and normal stress concentrations in the adhesive layer of plated RC beams. The Structural Engineer, Vol 67, No 12 / 20, June 1989, pp 229-233.

7. Roberts T M. Shear and normal stresses in adhesive joints. Journal of Engineering Mechanics, ASCE, Vol 115, No 11, Nov 1989, pp 2460-2479.

8. Al - Amery R I M and Roberts T M. Nonlinear finite difference analysis of composite beams with partial interaction. Computers and Structures, Vol 35, No 1, 1990, pp 81-87.

CH 4   SHEAR STRENGTH OF COMPOSITE PLATE GIRDERS WITH
WEB CUT OUTS

4.1   INTRODUCTION

In practice it is often necessary to provide openings in the webs of steel plate girders through which service ducts may pass, thereby reducing the overall construction depth (Figure 4.1). The presence of such openings may reduce significantly the shear strength of the girder, and necessitate the provision of expensive reinforcement around the hole to restore adequate strength. Alternatively it may be possible to take advantage of the composite action between the plate girder and associated concrete slab.

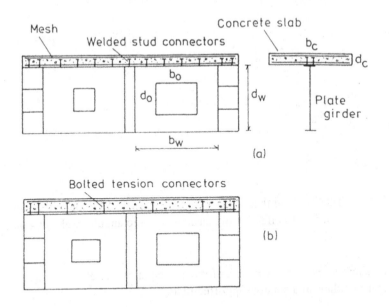

Figure 4.1   Composite plate girders with web perforations and (a) welded stud connectors (b) bolted tension connectors.

Experimental studies of the behaviour of short-span composite plate girders with web cut outs have indicated that web instability is an important factor in determining the shear strength of such girders, and that composite action can be beneficial when adequate connection is provided between the plate girder and concrete slab. The tests also indicated that it is the tensile or pull out capacity of the connectors that sustains the composite action under shear loading.

4.2   THEORY

4.2.1   Tension field theory

During the past thirty years the behaviour of slender plate girders loaded predominantly in shear has received much attention, and simple solutions based on assumed equilibrium stress fields have been developed for predicting ultimate loads. These solutions are based on the following assumptions, which are illustrated in Figure 4.2(a).

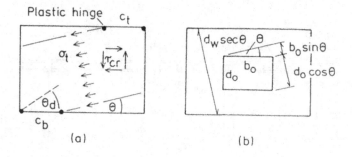

Figure 4.2   Tension field theory (a) assumed stress distribution in the web of a plate girder (b) effective dimensions of a rectangular web cut - out.

a)   The shear stress in the web is uniform and equal to the elastic critical stress τ_{cr} for shear buckling of a simply supported plate.

b)   After buckling a tension field of magnitude σ_t develops in the web, inclined at an angle θ to the flanges, which is related to τ_{cr} and the material yield stress σ_w via von Mises plastic yield criterion.

c)   Plastic hinges form in the top and bottom flanges, at distances c_t and c_b from the ends of the web panel, which are determined by considering the equilibrium of the flanges.

d)   The material is sufficiently ductile for the assumed stress fields to develop.

e)   End posts and adjacent panels are sufficiently strong to sustain the assumed stress fields.

Subject to assumptions a) d) and e), such solutions satisfy the requirements of a lower bound solution and maximise when the assumed inclination of the tension field θ is approximately two thirds of the inclination of the panel diagonal θ_d. The shear strength of the panel without cutouts V_g, can be determined from the following equations.

$$\tau_{cr} = \frac{k \, \pi^2 \, E \, t_w^2}{12 \, (1 - v^2) \, d_w^2} \tag{4.1}$$

$$k = 5.35 + 4 \, (d_w / b_w)^2 \qquad \text{when } b_w / d_w > 1 \tag{4.2}$$

$$k = 5.35 \, (d_w / b_w)^2 + 4 \qquad \text{when } b_w / d_w < 1 \tag{4.3}$$

$$\theta_d = \tan^{-1} (d_w / b_w) \tag{4.4}$$

$$\theta = 2 \, \theta_d / 3 \tag{4.5}$$

$$\sigma_t = -1.5 \, \tau_{cr} \sin2\theta + \{\sigma_w^2 + \tau_{cr}^2 \, (9 + 0.25 \sin^2 2\theta - 3)\}^{0.5} \tag{4.6}$$

$$M_{pt} = \sigma_{ft} \, b_{ft} \, t_{ft}^2 / 4 \tag{4.7}$$

$$M_{pb} = \sigma_{fb} \, b_{fb} \, t_{fb}^2 / 4 \tag{4.8}$$

$$c_t = 2 \, \{M_{pt} / \sigma_t \, t_w\}^{0.5} / \sin\theta \tag{4.9}$$

$$c_b = 2 \, \{M_{pb} / \sigma_b \, t_b\}^{0.5} / \sin\theta \tag{4.10}$$

$$V_g = \tau_{cr} \, d_w \, t_w + (c_t + c_b) \, t_w \, \sigma_t \sin^2\theta$$

$$+ \, \sigma_t \, t_w \, d_w \, (\cot\theta - \cot\theta_d) \sin^2\theta \tag{4.11}$$

4.2.2   Plate girders with web cut outs

Experimental studies have shown that the shear strength of plate girders with centrally placed web cut - outs V_{gp}, can be approximated by a linear interpolation between the strength of the unperforated girder V_g and the Vierendeel strength of the flanges V_v. Hence :

$$V_{gp} = V_g - \alpha (V_g - V_v) \qquad (4.12)$$

in which α is the effective depth ratio of the cut - out and :

$$V_v = 2 (M_{pt} + M_{pb}) / b_w \qquad (4.13)$$

The effective depth ratio of a rectangular cut - out, width b_o and depth d_o, in a stress field inclined at an angle θ to the horizontal, as shown in Figure 4.2(b), is given by :

$$\alpha = (b_o \sin\theta + d_o \cos\theta) / (d_w \sec\theta) \qquad (4.14)$$

4.2.3   Composite plate girders with web cut outs

It is recognised that the interaction between a perforated plate girder and a composite concrete slab loaded in shear is complex and variable. It is assumed however that the shear strength of the composite plate girder V_{gpc} can be approximated simply by adding a component V_c representing the composite contribution of the concrete slab. Hence :

$$V_{gpc} = V_g - \alpha (V_g - V_v) + V_c \qquad (4.15)$$

Experimental studies have indicated that it is the tensile or pull-out capacity of the connectors, either over the end post or in the central region of the web panel, which sustains the composite action during typical shear failure modes. Therefore, V_c is taken as the lesser of the shear strength of the concrete slab V_{cs} and the tensile or pull - out capacity of a single or group of connectors T_c.

In accordance with BS 8110, the shear strength of the concrete slab is given by :

$$V_{cs} = 0.79 \, b_{ce} \, d_{ce} \, \{100 \, A_s / b_{ce} \, d_{ce}\}^{0.333} \{400 / d_{ce}\}^{0.25} \{\sigma_{cu} / 25\}^{0.333} \qquad (4.16)$$

in which the units are Newtons and millimetres and the compressive cube strength of the concrete σ_{cu} is limited to 40 N/mm^2. b_{ce} and d_{ce} are the effective width and depth of the concrete slab and A_s is the area of the longitudinal tensile reinforcement.

Based on the assumed mode of failure illustrated in Figure 4.3, the pull out capacity of a single or group of stud connectors, spaced across the width of the plate girder flange, is given by :

$$T_c = \{\pi (B + G \cot\varphi) + 2 S\} G \sigma_{ct} \cot\varphi \qquad (4.17)$$

in which σ_{ct} is the tensile strength of the concrete and φ is assumed equal to 45^0. For a single connector S is equal to zero, while for more than two connectors S is the distance between the outer connectors. The value of T_c should not exceed the ultimate tensile capacity of the connectors.

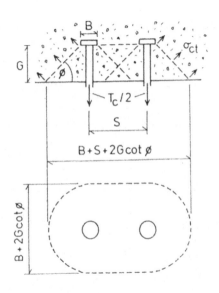

Figure 4.3   Theoretical model for predicting the pull - out capacity of stud connectors.

4.3   COMPARISON OF THEORETICAL AND EXPERIMENTAL RESULTS

Two series of tests were performed on small scale, short span, composite plate girders illustrated in Figure 4.1. Each girder consisted of two rectangular panels with central, rectangular web cut - outs, on top of which was cast a microconcrete slab, reinforced with two layers of 3.25 mm diameter mesh at 50 mm centres.

For the first series of twelve tests the concrete slabs were connected to the plate girders by pairs of 8 mm diameter headed studs, welded to the top flanges. For the second series of ten tests the concrete slabs were connected to the plate girders by pairs of 8 mm diameter bolts, which were located in oversized holes in the top flanges of the girders and therefore offered negligible resistance to horizontal shear forces.

The test results are presented graphically in Figure 4.4. For all the test girders the simple analytical model presented in previous sections provided a reasonably accurate and conservative estimate of the ultimate shear load (ratio of V_{ex} / V_{gpc} varying between 1.00 and 1.26). Three of the test girders exhibited pull out failure of the connectors, which was consistent with the theoretical predictions ($T_c < V_{cs}$).

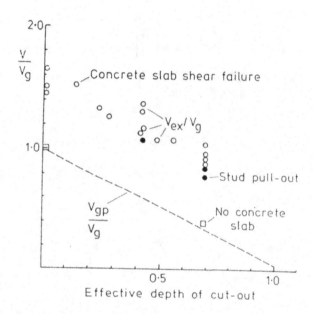

Figure 4.4   Comparison of experimental and theoretical shear strengths of composite plate girders.

4.4   NOTATION

A	area
b	width
B	diameter of head of stud
c	plastic hinge distance
d	depth
E	Young's modulus
G	length of shank of stud connector
k	shear buckling coefficient
M	moment
S	transverse spacing of stud connectors
t	thickness
T	ultimate tensile or pull out force
V	shear force
α	effective depth ratio of cut - out
θ	inclination of tension field
ν	Poisson's ratio
σ	material yield stress
τ	shear stress

Subscripts

b	bottom
c	concrete ; connectors
cr	critical
cs	concrete slab
ct	concrete tensile
cu	concrete cube compression
d	web panel diagonal
e	effective
ex	experimental
f	flange
g	girder
gp	perforated girder
gpc	composite perforated girder
o	cut - out
p	plastic
s	steel reinforcement
t	tension field ; top
w	web

4.5   BIBLIOGRAPHY

1.   Narayanan R, Al-Amery R I M and Roberts T M.   Shear strength of composite
      plate girders with rectangular web cut - outs.   J Construct Steel Research, 12,
      1989, pp 151-166.

2.   Roberts T M and Al-Amery R I M.   Shear strength of composite plate girders with
      web cut - outs.   Journal of Structural Engineering. ASCE, Vol 117, No 7, 1991,
      pp 1897-1910.

3.   BS 8100 : Part 1 : Structural use of concrete.   British Standards Institution,
      London, 1985.

4.   Porter D M, Rockey K C and Evans H R.   The collapse behaviour of plate girders
      loaded in shear.   The Structural Engineer, Vol 53, 1975, pp 313-325.

5.   Lawson R M.   Design for openings in the webs of composite beams.   SCI
      Publication 068, The Steel Construction Institute, Ascot, UK, 1987.

CH 5   COMPOSITE SLABS WITH PROFILED STEEL SHEETING

5.1   INTRODUCTION

Composite slabs with profiled steel sheeting (Figure 5.1) have been used in the construction of buildings, for many years in North America and more recently in parts of Europe. One of the main advantages of such slabs is the ease and speed of construction during which the profiled sheeting supports the weight of the wet concrete and the construction loads, thus enabling work to continue beneath. Once the concrete has hardened the profiled sheeting acts as tensile reinforcement, and the corrugations provide convenient ducts for incorporating services within the slab.

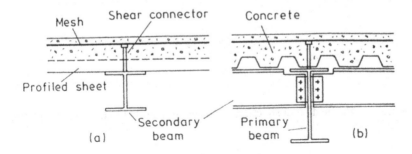

Figure 5.1   Composite floor system using profiled steel sheeting (a) composite slab supported by a secondary beam (b) secondary beams supported by a primary beam.

A typical floor system incorporating profiled steel sheeting, is illustrated in Figure 5.1. In general, due to span limitations (typically between 2.7 and 3.6m) the sheeting is laid at right angles to and is supported by the secondary beams, which in turn transfer the dead and imposed loads to the primary beams. The sheeting may be either simply supported or continuous over the secondary beams and is generally connected to the secondary and primary beams by stud connectors, welded through the troughs of the sheeting to the top flanges.

Modern deck profiles are made from galvanised steel sheet, typically 0.9 to 1.5mm thick. Profile heights are generally between 45 and 75 mm with a trough spacing between 150 and 300 mm. There are two well known generic types, namely, the dovetail profile and and the trapezoidal profile with web indentations (Figure 5.2). Dovetail profiles achieve composite action with the hardened concrete primarily by preventing separation. Trapezoidal profiles generally have web indentations or embossments, protruding into the hardened concrete, which provide enhanced mechanical interlocking.

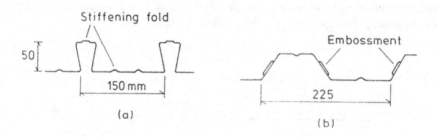

Figure 5.2   Profiled steel sheeting (a) dovetail profile (b) trapezoidal profile.

Concrete depths depend largely upon fire resistance and are normally between 100 and 150 mm (60 to 120 mm above the profile). Both normal and lightweight concrete are used. Conventional reinforcement is incorporated within the concrete, to control cracking and provide adequate bending resistance normal to the line of the corrugations.

5.2   PROFILED SHEETING AS PERMANENT FORMWORK

5.2.1   Design considerations

During construction, the profiled steel sheeting is required to support the weight of the wet concrete and the associated construction loads.

According to BS5950 : Part 4, the design strength of continuous decking should be based on an elastic distribution of moment as a safe lower bound. The bending strength of the section should also be determined elastically, since in general it will be limited by local buckling of the compression elements, either at mid span or over the supports. The strength of the section is improved by stiffening folds in the compression elements.

A limit on the residual deflection of the profile after concreting of span / 180 is also specified. This may be increased to span / 130 if the effects of ponding (an increase in the depth of concrete due to deflection of the sheeting) is taken into account.

5.2.2   Deflections and ponding

The deflection of profiled steel sheeting under the weight of wet concrete can be determined using elastic analysis. If ponding is taken into account, the governing differential equation (per unit width of sheeting) for determining the deflection (Figure 5.3) is :

$$E\,I\,\frac{d^4 w}{dx^4} = (d_c + w)\,\rho_c \qquad\qquad (5.1)$$

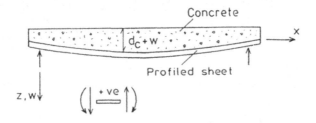

Figure 5.3   Ponding of profiled steel sheeting under the weight of wet concrete.

In equation (5.1) w is the deflection, E is Young's modulus, I is the second moment of area of the cross - section about its centroidal axis, d_c is the mean depth of the wet concrete excluding ponding and ρ_c is the unit weight of the wet concrete. The general solution of equation (5.2) is :

$$w = A\cos kx + B\sin kx + C\cosh kx + D\sinh kx - d_c \qquad\qquad (5.2)$$

$$k = (\rho_c\,/\,E\,I)^{0.25} \qquad\qquad (5.3)$$

The constants of integration A, B, C and D have to be determined from the boundary conditions, which in general is a complex and laborious process. In principal however, the deflection of the sheeting, including the influence of ponding, can be determined from equation (5.2).

5.2.3 Propped construction

The deflection of profiled steel sheeting during construction can be reduced by temporary props, as illustrated in Figure 5.4. The prop force P due to placing of the wet concrete is determined from elastic analysis. For simply supported sheeting of span 2L, supporting a uniformly distributed load $d_c \rho_c$ per unit length, the central prop force P and hogging moment M per unit width, are given by :

$$P \quad = \quad 1.25 \, d_c \, \rho_c \, L \tag{5.4}$$

$$M \quad = \quad 0.125 \, d_c \, \rho_c \, L^2 \tag{5.5}$$

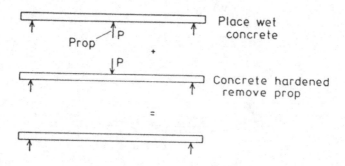

Figure 5.4   Propped construction.

When the concrete has hardened the prop is removed, which is equivalent to applying an equal and opposite force to the composite slab. The central deflection of the composite slab w_c, due to application of the prop force, can also be determined from elastic analysis and is given by :

$$w_c \quad = \quad P \, L^3 / (6 \, E \, I) \tag{5.6}$$

in which E I is the flexural rigidity of the transformed composite section, allowing for the cracking of concrete in tension.

The prop force should also be considered when assessing the strength of the composite slab.

5.2.4   Design by testing

The design of single span sheeting to support the weight of the wet concrete is generally governed by deflection limitations, while the design of continuous sheeting is often governed by strength criteria. Elastic analysis of simply supported sheeting does not take account of the beneficial influence of end restraint provided by the stud connectors, and elastic predictions of strength governed by local buckling are generally over conservative. Consequently, most manufacturers provide safe load - span tables, based predominantly if not entirely on test results (Figure 5.5(a)).

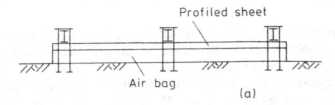

Profiled sheet

Air bag   (a)

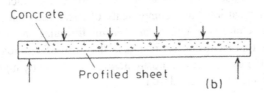

Concrete

Profiled sheet   (b)

Figure 5.5   Design by testing (a) air bag test on profiled steel sheeting
(b) load test on a composite slab.

5.3   COMPOSITE SLABS

5.3.1   Design considerations

Composite slabs are usually designed as simply supported elements, with no account taken of the continuity over supports provided by the conventional slab reinforcement. The ultimate bending capacity is generally controlled by slip between the concrete and steel sheeting. This occurs when the interface shearing forces exceed the chemical, frictional and mechanical interlocking resistances, and is known as shear bond failure.

If adequate end anchorage is provided, a composite slab may reach its ultimate bending capacity as a reinforced concrete slab, in which the decking acts as conventional tensile reinforcement.

The transverse shear resistance and serviceability limit states (control of premature shear bond failure and deflections) of composite slabs, should also be assessed.

5.3.2   Design by testing

The performance of a composite slab is influenced to a significant extent by its shear bond capacity, which can not be determined theoretically. Consequently, the design of such slabs is generally based on tests of the form shown in Figure 5.5(b). The slab is first subjected to dynamic loading between 50 and 150% of the desired working load, followed by a quasi - static loading to failure. The purpose of the dynamic loading is to identify slabs with an inherently fragile shear connection.

The test results are used to establish empirical constants which broadly define the friction bond and mechanical interlock components of shear resistance for the particular profile. Due to the empirical nature of design formulae in codes of practice, manufacturers normally represent the test results in the form of safe load - span tables. It is generally found that the degree of composite action is sufficiently high for most designs to be governed by the construction condition.

5.4   COMPOSITE BEAMS

Profiled steel sheeting is generally attached to the secondary and primary steel beams by stud connectors, welded through the steel plate onto the top flanges (Figure 5.1). An effective width of the composite slab can therefore be assumed to act compositely with the steel beams, thereby significantly increasing their bending resistance. The design of composite beams with profiled steel sheeting follows conventional procedures, with minor modifications to allow for the direction of the corrugations and their influence upon the shear strength of the connection.

5.5 NOTATION

d_c	mean depth of concrete
E	Young's modulus
I	second moment of area
L	dimension defining span
M	moment at prop location
P	prop force
w	displacement in z - direction
w_c	central deflection
x , z	coordinate axes

5.6 BIBLIOGRAPHY

1. BS5950 : The structural use of steelwork in buildings : Part 4 : Code of practice for design of floors with profiled steel sheeting. British Standards Institution, 1982.

2. Eurocode No. 4 : Design of composite steel and concrete structures : Part 1.1 : General rules and rules for buildings. 1992.

3. Lawson R M. Design of composite slabs and beams with steel decking. SCI Publication 055, The Steel Construction Institute, Ascot, UK, 1989.

CH 6   DOUBLE SKIN COMPOSITE CONSTRUCTION

6.1   INTRODUCTION

Double skin composite (DSC) or steel - concrete - steel sandwich (SCSS) construction is a new form of construction consisting of a layer of unreinforced concrete, sandwiched between two layers of relatively thin steel plate, connected to the concrete by welded stud shear connectors (Figure 6.1). The perceived advantages of DSC construction are that the external steel plates act as both permanent formwork and reinforcement and also as semi - impermeable, impact and blast resistant membranes. The overlapping welded stud connectors transfer shear and normal forces between the concrete and steel plates, and also act as transverse shear reinforcement, similar to the links in conventional reinforced concrete construction.

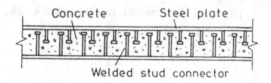

Figure 6.1   Double skin composite beam.

Double skin composite construction was developed initially as an alternative form of construction for immersed tube tunnels, but has since been considered for a variety of applications including offshore oil platforms, floating oil production and storage vessels, caissons, core shear walls in tall buildings and impact and blast resistant structures. Experimental and theoretical studies carried out to date have indicated that DSC elements and structures can generally be designed in accordance with conventional theories for reinforced concrete and composite steel and concrete structures, provided due consideration is given to the following modes of failure :

a)   crushing of concrete in compression
b)   yielding of the tension plate
c)   yielding and buckling of the compression plate
d)   slip yielding of the tension and compression plate connectors
e)   transverse shear failure
f)   tensile and pull out failure of the connectors at internal junctions subjected to opening forces.

6.2   STRENGTH OF SECTIONS IN BENDING

The ultimate moment of resistance of a DSC beam can be determined by considering fully plastic stress blocks in the concrete and steel plate, as illustrated in Figure 6.2. Concrete below the neutral axis is assumed to be cracked and does not therefore contribute to the strength of the section.

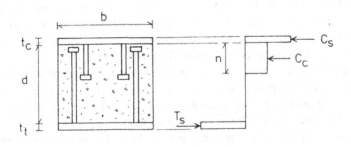

Figure 6.2   Bending of a double skin composite beam.

Assuming adequate shear connection between the concrete and steel plates, the ultimate tensile force T_s and compressive forces C_c and C_s are given by :

$$T_s = f_{sk} \, b \, t_t / \gamma_s \tag{6.1}$$

$$C_c = 0.67 \, f_{cu} \, b \, n / \gamma_c \tag{6.2}$$

$$C_s = f_{sk} \, b \, tc / \gamma_s \tag{6.3}$$

Equating the resultant axial force to zero gives the depth of the plastic neutral axis n as :

$$n = \frac{(T_s - C_s) \gamma_c}{0.67 \, f_{cu} \, b} \qquad (6.4)$$

Taking moments about the line of action of T_s gives the ultimate resistance moment M_{Rd} as :

$$M_{Rd} = C_s (d + t_t / 2 + t_c / 2) + C_c (d + t_t / 2 - n / 2) \qquad (6.5)$$

6.3   STRENGTH OF SECTIONS IN SHEAR

The ultimate transverse shear resistance of a DSC element is provided by the concrete and the full depth or overlapping stud connectors. On the basis of numerous tests, it appears that the clauses given in BS8110 for calculating transverse shear resistance, can be applied with some modifications to DSC elements.

The concrete in DSC elements subjected to bending may have discrete cracks emanating from the stud connectors, which are likely to reduce the basic shear strength of the concrete. The basic values of concrete shear strength given in BS8110 should therefore be reduced by 20%. For determining these values, the area of tensile reinforcement should be taken as the area of the tension plate.

Stud connectors assumed to act as transverse shear reinforcement, should be welded to the tension plate and extend well into the concrete compression zone (Figure 6.2). Such studs should preferably extend to the full depth of the section, so that they act as spacers during fabrication. Where full depth studs are not employed, they should be overlapped with studs welded to the compression plate. The maximum longitudinal spacing of studs assumed to act as transverse shear reinforcement, should be limited to 0.75 of the concrete depth.

6.4   STRENGTH OF SECTIONS SUBJECTED TO BENDING AND AXIAL FORCES

The resistance of a DSC element to both bending and axial forces, can be determined in accordance with conventional theories for doubly reinforced concrete sections. Assuming plane sections remain plane and that the concrete is subjected to a limiting compressive strain $\varepsilon_c = 0.0035$, the assumed strain and plastic stress distributions are as shown in Figure 6.3(a). The design axial compressive force P_d and moment M_d are assumed to act at the centroid of the uncracked transformed section.

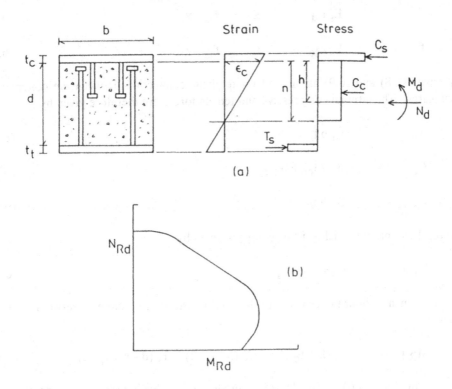

b

Strain

Stress

(a)

(b)

Figure 6.3   Double skin composite section subjected to bending and axial force
(a) strain and stress profiles (b) interaction diagram
for axial and bending resistances.

From the assumed linear strain distribution, the average strains in the tension and
compression plates ε_{st} and ε_{sc}, are given by :

$$\varepsilon_{st} \quad = \quad \varepsilon_c \, (d - n + t_t \, / \, 2) \, / \, n \tag{6.6}$$

$$\varepsilon_{sc} \quad = \quad \varepsilon_c \, (n + t_c \, / \, 2) \, / \, n \tag{6.7}$$

Stresses in the tension and compression plates f_{st} and f_{sc}, are therefore :

$$f_{st} = E_s \varepsilon_{st} \leq f_{sk} / \gamma_s \tag{6.8}$$

$$f_{sc} = E_s \varepsilon_{sc} \leq f_{sk} / \gamma_s \tag{6.9}$$

Equations (6.8) and (6.9) assume adequate shear connection between the concrete and steel plates. The internal compressive and tensile forces are therefore given by :

$$T_s = f_{st} b t_t \tag{6.10}$$

$$C_c = 0.67 f_{cu} b n / \gamma_c \tag{6.11}$$

$$C_s = f_{sc} b t_c \tag{6.12}$$

The resultant internal axial resistance N_{Rd} is given by :

$$N_{Rd} = C_s + C_c - T_s \tag{6.13}$$

Taking moments about the centroid of the section, the internal moment resistance M_{Rd} is given by :

$$M_{Rd} = C_s (h + t_c / 2) + C_c (h - n / 2) + T_s (d - h + t_t / 2) \tag{6.14}$$

Equations (6.6) to (6.14) are generally solved by iteration. The depth of the plastic neutral axis is assumed, which enables N_{Rd} and M_{Rd} to be determined and compared with the design axial force N_d and moment M_d. The resistances of the section are considered to be adequate if a value of n exists for which :

$$N_{Rd} \geq N_d \tag{6.15}$$

$$M_{Rd} \geq M_d \tag{6.16}$$

Alternatively, an interaction diagram can be plotted representing values of N_{Rd} and M_{Rd} for all possible values of n between zero and the concrete depth d (Figure 6.3(b)). If N_d and M_d lie within the interaction diagram, the resistances of the section are adequate.

6.5   SHEAR CONNECTION AND COMPRESSION PLATE BUCKLING

The equations presented in sections 6.2 and 6.4 are based on the assumption that sufficient shear connectors are provided to resist the ultimate tensile and compressive forces in the external steel plates. Experimental evidence indicates that the performance of stud connectors in concrete tension zones is less favourable than in concrete compression zones. Consequently, it has been recommended that the ultimate forces in the tension and compression plates are also limited by the equations

$$T_s \leq 0.5\, n_t\, P_{Rk} \qquad\qquad (6.17)$$

$$C_s \leq 0.8\, n_c\, P_{Rk} \qquad\qquad (6.18)$$

in which n_t and n_c are the number of connectors effective in resisting the axial forces in the tension and compression plates and P_{Rk} is the characteristic resistance of the connectors. It is not clear however whether or not the factors 0.5 and 0.8 are intended to include the partial material safely factor for studs γ_v, which in Eurocode 4 has been specified as 1.25.

Equations (6.17) and (6.18) can be used as a basis for either full connection design, which is governed by yielding of the steel plates, or partial connection design, which is governed by slip yielding of the connectors.

Compression plate buckling has also been observed during a number of tests on DSC beams and columns. It has been recommended that the longitudinal spacing of the compression plate connectors be limited to 40 times the plate thicknesss, to avoid this phenomenon.

6.6   BENDING STIFFNESS OF DOUBLE SKIN COMPOSITE ELEMENTS

Experimental studies have indicated that the flexural stiffness of DSC elements, in particular with partial shear connection, is likely to be influenced to a significant extent by slip between the concrete and external steel plates. Theoretical predictions based on partial interaction theory show reasonable correlation with test results, but are generally too complicated for routine design. However, approximate solutions can be obtained by considering reduced effective areas of steel plate, based on the degree of shear connection, to act compositely with the cracked concrete section.

6.7   ACKNOWLEDGEMENTS

This chapter is based primarily upon research carried out in the School of Engineering at the University of Wales Cardiff, and sponsored by the Science and Engineering Research Council, UK. Current research in the School is being sponsored by the European Commission and British Steel, and coordinated by the Steel Construction Institute.

6.8   NOTATION

b	width of section
C	compressive force
d	depth of concrete
E	Young's modulus
f	stress
h	depth of centroid from the surface of the concrete
M	moment
n	depth of plastic neutral axis from the surface of the concrete ; number of effective shear connectors
N	axial force
P	connector shear force
t	steel plate thickness
T	Tensile force
γ	partial safety factor for materials
ε	strain

Subscripts

c	concrete ; compression
cu	concrete cube compression
d	design
k	characteristic
R	resistance
s	steel
t	tension
v	stud connectors

6.9   BIBLIOGRAPHY

1.   Tomlinson M, Chapman M, Wright H D, Tomlinson A and Jefferson A. Shell composite construction for shallow draft immersed tube tunnels. ICE International Conference on Immersed Tube Tunnel Techniques, Manchester, UK, 1989, pp 209-220.

2.   Wright H D, Oduyemi T O S and Evans H R. The experimental behaviour of double skin composite elements. J Construct Steel Research, 19, 1991, pp 97-110.

3.   Wright H D, Oduyemi T O S and Evans H R. The design of double skin composite elements. J Construct Steel Research, 19, 1991, pp 111-132.

4.   Wright H D and Oduyemi T O S. Partial interaction analysis of double skin composite beams. J Construct Steel Research, 19, 1991, pp 253-283.

5.   Narayanan R, Lee I L, Roberts T M, Evans H R, Edwards D N and Helou A J. Studies on steel concrete steel sandwich construction. Options for Tunnelling 1993, Ed H Berger, Elsevier Science Publishers, Amsterdam, 1993, pp 193-202.

6.   Narayanan R and Roberts T M. Double skin composite construction for submerged tube tunnels. Composite Construction in Steel and Concrete II, Eds W S Easterling and W M K Roddis, ASCE, New York, 1993, pp 351-365.

7.   Narayanan R, Roberts T M and Naji F J. Design guide for steel-concrete-steel sandwich construction. The Steel Construction Institute, Ascot, UK, 1994.

8.   BS 8110 : Structural use of concrete. British Standards Institution, 1985.

9.   Eurocode No 4 : Design of composite steel and concrete structures : Part 1.1 : General rules and rules for buildings, 1992.

Printed in the United States
By Bookmasters